Demilitarisation and International Law in Context

This book calls our attention to a pioneering demilitarization agreement in the 19th century, that contributes greatly to our understanding of the larger phenomenon. A splendid example of the utility of a vigorous case study.
–Professor Peter Stearns

The demilitarisation and neutralisation of the Åland Islands is a confirmation of, and an exception to, the collective security system in present-day international affairs. Its core idea is that there is no need for military presence in the territory of the islands and that they are to be kept out of military activities. A restricted use of military force has a confidence building effect in cases where competing interests may be so intense that banning the very presence of military force remains the only viable option. The regime of the Åland Islands is the result of pragmatic and contingent political compromises. As such, the case of the Åland Islands offers an alternative trajectory to the increased militarisation we witness around the world today.

Through parliamentary and archival materials, international treaties and academic works, the authors examine the legal rules and institutional structures of the demilitarisation regime. In this process they reassess core concepts of international law and international affairs, such as sovereignty and security, and introduce a theoretical view on the empirical case study of the Åland Islands. The book covers legal, political and policy discursive aspects of demilitarisation, international co-operation, defence and security matters around the Baltic Sea with a broader European and global relevance. It can be a source of inspiration for all those in search of constructive efforts that can address territorial disputes and security challenges.

Sia Spiliopoulou Åkermark, Jur. dr., Associate Professor (docent) in international law, Director, The Åland Islands Peace Institute (Finland). Her work focuses on diversity, autonomy, the use of force and international law and its institutions.

Saila Heinikoski, Ph.D. (Political Science), M.A. (University of Turku, Finland). Her work focuses on the European Union, European and national foreign and security policies, mobility and political discourses.

Pirjo Kleemola-Juntunen, LL.D., LL.Lic., LL.M., Postdoctoral Researcher, Northern Institute for Minority and Environmental Law/Artic Centre, University of Lapland. Her work focuses on international law in particular the law of the sea and international environmental law.

Routledge Research in International Law

Available:

Means of Transportation and Registration of Nationality
Transportation Registered by International Organizations
Vincent P. Cogliati-Bantz

Regionalism in International Law
Ján Klučka

The International Criminal Court and Nigeria
Implementing the Complementarity Principle of the Rome Statute
Muyiwa Adigun

Armed Conflict and Forcible Displacement
Individual Rights under International Law
Elena Katselli Proukaki

The Rule of Unwritten International Law
Customary Law, General Principles, and World Order
Peter G. Staubach

State Interest and the Sources of International Law
Doctrine, Morality, and Non-Treaty Law
Markus P. Beham

Demilitarisation and International Law in Context
The Åland Islands
Sia Spiliopoulou Åkermark, Saila Heinikoski and Pirjo Kleemola-Juntunen

For a full list of titles in this series, visit www.routledge.com/Routledge-Research-in-International-Law/book-series/INTNLLAW

Demilitarisation and International Law in Context

The Åland Islands

Sia Spiliopoulou Åkermark, Saila Heinikoski and Pirjo Kleemola-Juntunen

Routledge
Taylor & Francis Group
NEW YORK AND LONDON

First published 2018
by Routledge
711 Third Avenue, New York, NY 10017

and by Routledge
2 Park Square, Milton Park, Abingdon, Oxon OX14 4RN

Routledge is an imprint of the Taylor & Francis Group, an informa business

British Library Cataloguing-in-Publication Data
A catalogue record for this book is available from the British Library

Library of Congress Cataloging-in-Publication Data
Names: Spiliopoulou Åkermark, Athanasia, author. | Heinikoski, Saila | Kleemola-Juntunen, Pirjo.
Title: Demilitarisation and international law in context : the Åland islands / Sia Spiliopoulou Åkermark, Saila Heinikoski and Pirjo Kleemola-Juntunen.
Description: Abingdon, Oxon ; New York, NY : Routledge, 2018. | Includes bibliographical references and index.
Identifiers: LCCN 2018003465 | ISBN 9781138093300 (hardback)
Subjects: LCSH: Åland (Finland)—International status. | Neutrality—Finland—Åland. | Disarmament—Finland—Åland. | Islands—Law and legislation—Finland—Åland.
Classification: LCC KZ4214 .S67 2018 | DDC 341.4/20948972—dc23
LC record available at https://lccn.loc.gov/2018003465

ISBN: 978-1-138-09330-0 (hbk)
ISBN: 978-1-315-10690-8 (ebk)

Typeset in Times New Roman
by Apex CoVantage, LLC

Contents

Foreword

We have been fortunate to be given the chance to work on the research project "Demilitarisation in an increasingly militarised world—International perspectives in a multilevel framework: the case of the Åland Islands". The project was funded by the Kone Foundation but has also received a small, but important, contribution from the Åland Culture Foundation, *Ålands kulturstiftelse*. The generosity and unbureaucratic ethos of both those institutions have been crucial in facilitating the co-operation between the two partners involved, namely the Åland Islands Peace Institute in Mariehamn and the Nordic and Environmental Law Institute/Arctic Centre, University of Lapland in Rovaniemi. Our colleagues have offered warm support and constructive criticism for which we are grateful.

We were strengthened in our endeavour by the fact that the project was initially approved as being a 'bold' enterprise. The challenge has been to contextualise the demilitarised experience within broader global trends of emphasis on security and re-armament, debates on the role of law domestically and internationally, as well as within efforts for deepened European integration. This is a new way of looking at the wider importance of the demilitarisation and neutralisation of the Åland Islands.

This book is co-authored, combining knowledge and inspiration from several scientific fields. The fact that it is co-authored means that the authors share the core arguments made. Sia holds prime responsibility for Chapter 2, Pirjo for Chapter 3 and Saila for most parts of Chapter 4. Chapters 1 and 5 have been a process of true co-authorship between Sia and Saila.

The outcome of the project includes many more articles, book chapters and debate articles written in the past almost three years by the three authors and by our wonderful colleagues during various stages of the project. Timo Koivurova, Filip Holiençin and Yannick Poullie have all made strong contributions to this work and to the richness and pleasure of our common discussions. The articles are included in the literature list of the book.

The research project has benefited from the knowledge of its interdisciplinary scientific board. Its composition included Matthieu Chillaud, Lauri Hannikainen, Kenneth Gustavsson, Päivi Kaukoranta, Said Mahmoudi, Allan Rosas, Gregory Simons and Geir Ulfstein. Alyson Bailes, whose loss was an unexpected blow, was succeeded by Willy Østreng, to whom we are warmly indebted. It is a great pity that Alyson is not here with us to discuss and enjoy the fruits of this co-operative work. Special thanks go to Said Mahmoudi and Gunilla Herolf (member of the scientific board of the Åland Islands Peace Institute) who have commented on various drafts of the book and to Lauri Hannikainen for his consistent encouragement. We thank professor emeritus Peter Stearns for his support in the final steps of the publication process.

The warmest expression of gratitude goes, as always, to our partners and families.

Sia, Saila and Pirjo
In Mariehamn, Lappeenranta and Rovaniemi
December 2017

Co-authors

Sia Spiliopoulou Åkermark, Jur. dr., Associate Professor (docent) in international law, Director, The Åland Islands Peace Institute (Finland). Her work focuses on diversity, autonomy, the use of force and international law and its institutions.

Saila Heinikoski, Ph.D. (Political Science), M.A. (University of Turku, Finland). Her work focuses on the European Union, European and national foreign and security policies, mobility and political discourses.

Pirjo Kleemola-Juntunen, LL.D., LL.Lic., LL.M., Postdoctoral Researcher, Northern Institute for Minority and Environmental Law/Artic Centre, University of Lapland. Her work focuses on international law in particular the law of the sea and international environmental law.

Abbreviations

BALTOPS	Baltic Operations (military exercise)
BRIC	Brazil, Russia, India and China
CBSS	Council of the Baltic Sea States
CFSP	Common Foreign and Security Policy of the European Union
CIMIC	Civil-military interaction
CISE	Common Information Sharing Environment
CSDP	Common Security and Defence Policy of the European Union
COREPER	Committee of the Permanent Representatives of the Governments of the Member States to the European Union
DoD	Department of Defense (United States)
EEZ	Exclusive Economic Zone
ESDP	European Security and Defence Policy (later CSDP)
EU	European Union
EUSBR	European Union Strategy for the Baltic Sea Region
FCO	Foreign and Commonwealth Office (Great Britain)
FTS	Finnish Treaty Series
HELCOM	Helsinki Commission
ICJ	International Court of Justice
LNTS	League of Nations Treaty Series
LoN	League of Nations
LOSB	Law of the Sea Bulletin
MARSUR	Maritime Surveillance Networking
MoU	Memorandum of Understanding
MP	Member of Parliament
NATO	North Atlantic Treaty Organization
NB6	Nordic-Baltic six cooperation (within the European Union)
NB8	Nordic-Baltic eight cooperation

nm	Nautical mile
NEFAB	North European Functional Airspace Block
NORDAC	Nordic Armaments Cooperation
NORDCAPS	Nordic Coordinated Arrangement for Military Peace Support
NORDEFCO	Nordic Defence Cooperation
NORDSUP	Nordic Supportive Defence Structures
OSCE	Organization for Security and Co-operation in Europe
PESCO	Permanent Structured Cooperation in defence within the EU
PSI	Proliferation Security Initiative
SOP	Statement of Interdiction Principles
SUCBAS	Sea Surveillance Co-operation Baltic Sea
SUCFIS	Sea Surveillance Co-operation Finland–Sweden
TSC	Territorial Sea and the Contiguous Zone
TEU	Treaty of the European Union (Lisbon Treaty)
UK	United Kingdom
UN	United Nations
UNCLOS	United Nations Conference on the Law of the Sea
LOSC	Law of the Sea Convention (1982)
OCHA	United Nations Office for the Coordination of Humanitarian Affairs
RP or HE	Government Bill in Finnish Parliament (*Riksdagens proposition*, in Swedish or *Hallituksen esitys* in Finnish)
SYKE	Finnish Environment Institute
TFEU	Treaty on the Functioning of the EU
UNTS	United Nations Treaty Series
USSR	Union of Soviet Socialist Republics
US	United States
WEU	Western European Union

1 Introduction

The Goal and Structure of the Book

The goal of the book is to look at the demilitarisation and neutralisation of Åland as a long-standing regime intended to limit war. International law has had a pacifist ambition from its early days dating back at least to Grotius' *De Jure Belli ac Pacis* and ranging from limitations of a perceived right to wage war and the use of military force, to limitations in warfare itself, in particular through the development of humanitarian law.[1] The regime of the Åland Islands has evolved over more than 160 years of demilitarisation and 100 years of neutralisation and has withstood several periods of test and change, including, most notably two world wars, the process towards the independence of Finland, the Russian revolution, the Cold War and its aftermath, and Finland's entry in the European Union (EU). This alone deserves attention and reflection. New methods of security and defence co-operation and of contestation as well as technological developments with regard to weapons, surveillance, communication and transportation prompt new questions regarding this evolving regime. These are the questions which will be dealt with in the present book. Throughout the book the reader will be introduced to combinations and reconfigurations of realist and idealist strands in international law and relations. It is not without reason that the word 'entanglement' is so frequently used in social sciences and humanities today. The comforting sharp polarities of modernity are now under reassessment.

The method employed is that of identifying and analysing key moments, documents and debates concerning the evolution of demilitarisation as it has been shaped and used in the case of the Åland Islands. International law is examined in the context not only of Finland as the state holding sovereignty over the islands, but also in the context of the Baltic Sea region, the

1 H. Grotius, *Jure Belli Ac Pacis Libri Tres* [On the Law of War and Peace: Three Books] (1646/1925).

role of neighbouring countries, European and Nordic processes of regional integration, divergence and convergence.

Any text is written at a particular moment of time and must address concerns of that specific time. The authors are interested in how the demilitarisation and neutralisation regime of the Åland Islands is affected by the development of modern technology and by the proliferation and varied modes of military-civilian co-operation in Europe and the Baltic Sea. Technology and co-operation are developed within an increased fusion between civil and military activities, civil and military concepts, norms and institutions as well as an increased discursive emphasis on security, threats and armed force.[2]

The effects of new technology on the development of international law, including in areas such as the law of the sea, telecommunications, use of weapons and aviation, is of course nothing new. Technological development had to be accommodated, for example regarding wireless telegraphy, prompting the 1907 Hague Conference to include several such provisions in Convention V on the Rights and Duties of Neutral Powers and Persons (Articles 3, 8 and 9).[3] In our research we have looked at the effects of technology on the demilitarisation regime.[4]

The present book starts with the introduction (Chapter 1) and touches upon some of the crucial concepts and assumptions of demilitarisation and neutralisation. Thereafter follows an overview of the legal framework pertaining to the Åland demilitarisation and neutralisation and raises questions about its continuity, its oversight and about the effects of new technology (Chapter 2). Chapter 3 is dedicated to the law of the sea rules and considerations affecting the Åland Islands, in particular issues of rights of passage. Even though technology and civil-military co-operation affect the demilitarisation regime not only at sea but in many different fields, it is still the case that more than 15% of world cargo traffic travels through the Baltic Sea. The Russian Federation holds the largest market share with Sweden close thereafter.[5] Trade interests therefore exist alongside security priorities in the Baltic Sea. Finally, Chapter 4 looks at various forms of security

2 U. Beck, *World Risk Society* (1999).

3 J.H.W. Verzijl, *International Law in Historical Perspective. Volume 10, The Law of Neutrality* (1979) 210.

4 S. Spiliopoulou Åkermark, "The Meaning of Airspace Sovereignty Today—A Case Study on Demilitarisation and Functional Airspace Blocks," *Nordic Journal of International Law* 86, No. 1 (2017) 91–117. By the same author, "Old Rules and New Technology: Drones and the Demilitarisation and Neutralisation of the Åland Islands" (Forthcoming).

5 Helsinki Commission (HELCOM), "Ensuring Safe Shipping in the Baltic," 2009.

co-operation affecting the Åland Islands. Conclusions and a forward-looking assessment are included in the final chapter (Chapter 5).

In the present book, the aim is not to give the *longue durée* history of the Baltic Sea and the thousands of islands called the Åland Islands. The focus is only on one aspect of this history, namely their demilitarisation and neutralisation starting in the middle of the 19th century. The so-called 'Åland Example' with its component elements of minority protection, cultural guarantees and territorial political autonomy, coupled with a regime of demilitarisation and neutralisation, is quite well known among those interested in international affairs along with the international community's original involvement through the League of Nations in finding a solution to the sovereignty dispute over the islands. These experiences are seen primarily through the lens of self-determination or as tools for the solution of disputes over territory.[6] The demilitarisation regime as such, however, has not attracted similar attention in international scholarship over the past century.[7]

1.1 Early Occurrence of Demilitarisation

The 19th century saw the institutionalisation of international law, essentially the law between new nation-states and disintegrating empires, with the large and colonial powers of Europe at the core of the project. This process was pursued through various formalised agreements, including for the purpose of establishing borders and shaping conditions of war and peace.[8] The demilitarisation of a delimited territory was nothing new at that time. Examples appear to date back to the early Middle Ages and rules requiring the demolition of fortifications and prohibiting their reconstruction are found in peace treaties concluded in Europe in the 17th and 18th centuries. An early example was the 1559 Treaty of Cateau-Cambrésis between France and Spain, which included a prohibition to construct fortifications in

6 S. Spiliopoulou Åkermark (ed.), *The Åland Example and Its Components: Relevance for International Conflict Resolution* (2011); T. Modeen, *De folkrättsliga garantierna för bevarandet av Ålandsöarnas nationella karaktär* (1973); J. Barros, *The Aland Islands Question: Its Settlement by the League of Nations* (1968).

7 J.O. Söderhjelm, *Démilitarisation et neutralisation des Iles d'Aland en 1856 et 1921* (1928); M. Björkholm and A. Rosas, *Ålandsöarnas demilitarisering och neutralisering* (1990); L. Hannikainen and F. Horn, *Autonomy and Demilitarisation in International Law: The Åland Islands in a Changing Europe*, eds. L. Hannikainen and F. Horn (1997); P. Kleemola-Juntunen, *Passage Rights in International Law: A Case Study of the Territorial Waters of the Åland Islands* (2014).

8 M. Koskenniemi, *The Gentle Civilizer of Nations: The Rise and Fall of International Law 1870–1960* (2001).

the area of Thérouanne. The peace treaty between Spain and the Low Countries (Münster 1648) ordered the demolition of fortifications in the border regions of Flanders and along the Scheldt River. The treaty also contained a general prohibition against the establishment of military constructions and strategic canals in this region. In 1768, Denmark ceded several islands in the mouth of the Elbe River to Hamburg and it was provided that no military installations were to be built on these islands.[9]

The decline of the Ottoman Empire and the reconfiguration of the European political scene during the first half of the 19th century gave rise to a number of demilitarisation commitments. In December 1832, a few years after the proclamation of independence by Greece, Turkey declared that the country had no intention of stationing troops on the Island of Samos in the Aegean Sea. This declaration constitutes a rare example of a unilateral measure with legal consequences of demilitarisation and was later followed up by the 1923 Treaty of Lausanne.[10] It is striking that the notion of demilitarisation and neutralisation of territories has not attracted much attention in international law, international relations or other academic fields. The reasons for this may be multiple. Firstly, demilitarisation and neutralisation are conceptually perceived as an anomaly to the orthodox understanding of the full territorial control by states and their governments. The Montevideo Convention on the Rights and Duties of States tried to define the conditions and effects of states in 1933. The convention provided in its first article:

> The state as a person of international law should possess the following qualifications: a) a permanent population; b) a defined territory; c) government; and d) capacity to enter into relations with the other states.

The three first elements have their origins, at least as a cluster, in the 19th century writings of Jellinek,[11] and it has been argued that control of territory is the essence of the state.[12] Control of territory can, however, be exercised and asserted in many ways, especially today when modern surveillance technology at a distance, economic interdependence and means of transport and communications offer a huge potential of control not only by states but also by other actors. In former times, however, a crucial tool for such control was the presence or, at least, the possibility of the presence and effective

9 J.H.W. Verzijl, *International Law in Historical Perspective*, Volume 3, *State Territory* (1970); C. Ahlström, *Demilitarised and Neutralised Territories in Europe* (2004).

10 Verzijl, *International Law in Historical Perspective: Volume 3, State Territory* (1970) 505.

11 G. Jellinek, *Allgemeine Staatslehre*, 3rd edn. (1914) 394 ff.

12 P. Malanczuk and M.B. Akehurst, *Akehurst's Modern Introduction to International Law* (2002).

control of the state concerned by the military.[13] Consequently, allowing for the demilitarisation of a certain part of the territory was a strong deviation from the idea of full territorial control. This logic is reflected in theories of military vacuum, which were often invoked during the Cold War, but persist still today. From a military logic perspective, this means that a demilitarised region as a case of a 'military vacuum' is an 'anomaly' that, according to some observers, should be 'rectified' by the re-introduction of military presence.[14] This is why title to and control of territory has been a core concern for international law and relations. Less attention has been paid to situations where such control is layered, partial or otherwise modified.[15]

Another reason why demilitarisation in law and in politics is perceived as 'sensitive' may be that such regimes have often been imposed on the defeated side by the victors of wars and battles. This punitive character may be seen as giving the concerned state a 'pariah' stamp.[16] In a recent publication on demilitarisation, the concept is used by looking at country-wide policies, and illustrated by the cases of Germany and Japan after the Second World War, coupled by what could be described as US interventionism in Latin America, or the avoidance of it, especially during the Cold War.[17] The vision in operation was encapsulated in the Joint Statement by US President Roosevelt and Prime Minister Churchill on 14 August 1941, in which they assured the world that while their countries "seek no aggrandizement, territorial or other", they would ensure the "disarmament" of aggressor states at the end of the war.[18]

The logic of the demilitarisation was, and still is, that of ensuring that a delimited piece of territory, often a border territory or an island, will not be fortified and therefore would be less attractive militarily and less dangerous than it would be otherwise. Demilitarisation denotes the reduction or abolishment of armament and military presence in a certain geographic area. Kingma and Schrijver note that in many respects, demilitarisation is the

13 J. Castellino et al., *Title to Territory in International Law: A Temporal Analysis* (2003).

14 Military presence of a different nature figures prominently also in the *Islands of Palmas Case*, RIIA (1928), Vol. II, 829–71. See also comments by the Finnish Defence Minister J. Niinistö, "A Demilitarised Åland Is a Military Vacuum," *Helsinki Times*, 18 October 2016.

15 M. Koskenniemi, "Occupation and Sovereignty—Still a Useful Distinction?," in *Law at War: The Law as It Was and the Law as It Should Be*, eds. O. Engdahl and P. Wrange (2008) 163–74.

16 D.J. Bederman, "Collective Security, Demilitarization and 'Pariah' States," *European Journal of International Law* 13, No. 1 (2002) 121–38.

17 P.N. Stearns, *Demilitarization in the Contemporary World* (2013).

18 "The Atlantic Conference: Joint Statement by President Roosevelt and Prime Minister Churchill," 14 August 1941.

opposite of militarisation.[19] These authors define militarisation as "the complex processes of mobilization of resources for armed conflict at political, economic and social levels, with high level influence of the military in the society". Some elements of the opposite processes of demilitarisation are, according to the same authors, increased civilian control over armed forces and a reduced reliance on the threat or the use of force in conflict management alongside the actual reduction of military expenditure.

The need for democratic control of armed forces has preoccupied lawyers and political thinkers for a long time and has become one of the foundational principles of democracy as an institution, as we can conclude from the work of Weber (1918) and the Venice Commission for Democracy through Law (2008) alike.[20] Max Weber's 1918 essay appeared within a wider debate on militarism in Germany and globally in the early 20th century. In journals such as *The Advocate of Peace*, published in the US, there was similarly an intense debate on militarism.[21] The German writers of that time were concerned about the link between processes of militarisation and the outbreak of war. Weber, however, argued that in politics, there is always an element of *chance* and a possibility to have different outcomes through choices and action.

Militarism as a phenomenon or an ideology, and militarisation as a political, societal and legal process can be approached from different disciplinary, theoretical and methodological angles. Following Shaw, militarism is the "penetration of social relations in general by military relations", while militarisation is the extension of such militarism.[22] This concept can be used in an analytical and descriptive way or in a critical and even polemical manner. It is clear, however, that these concepts involve much more than the quantitative growth of armies or military budgets.[23] Their impact is visible in the content and procedures of politics and by consequence in the content of law and the legislative process. Militarism and militarisation affect the discourses within professional groups and networks, whether they are of a

19 K. Kingma and N. Schrijver, *Demilitarization, Max Planck Encyclopedia of Public International Law* (2013).

20 M. Weber et al., *Weber: Political Writings* (1994) in particular in the essay "Parliament and Government in Germany Under a New Political Order" (1918); Venice Commission (Council of Europe), "Report on the Democratic Control of the Armed Forces," 2008.

21 V.R. Berghahn, *Militarism: The History of an International Debate 1861–1979* (1981); A. Vagts, *A History of Militarism: Civilian and Military* (1967).

22 M. Shaw, "Twenty-First Century Militarism: A Historical-Sociological Framework," in *Militarism and International Relations: Political Economy, Security, Theory*, eds. A. Stavrianakis and J. Selby (2012) 19–32.

23 A. Stavrianakis, "Legitimising Liberal Militarism: Politics, Law and War in the Arms Trade Treaty," *Third World Quarterly* 37, No. 5 (2016) 840–65.

civilian or a military nature.[24] In the present work, the focus is not on militarisation and militarism as such, but rather on demilitarisation as a concept and tool employed in international law and affairs.

1.2 The Concept of Demilitarisation

It has been argued that the concept of demilitarisation can be employed in a wide range of cases, from the limitation of arms in occupied or defeated countries and regions, to broader and democratically entrenched reductions of military budgets and apparatus, or even changes in societal norms and practices, changing the "militarism of the mind".[25] These various facets of demilitarisation all have in common, however, the will and interest in "limiting war".[26] The contingently defined concept of peace is at the heart of such efforts and has taken the form of various expressions, ranging from a focus on free trade and economic development, to democracy, to international administration, to conflict resolution, to collective security and to disarmament and demilitarisation. There are of course many other possible epistemological approaches to the notion of demilitarisation as a sociological, historical, anthropological or political concept. Bickford notes, for instance, that "demilitarization programs are desired forms and visions of the state and the moral universe of the state".[27] In the present book, however, we shall limit our inquiry to the ways limited territorial demilitarisation based on a contractual basis and implemented over a long period of time—as opposed to imposed state-wide demilitarisation—operates in the particular case of the Åland Islands.[28]

This book focuses on the experiences and contingencies of the Baltic Sea region. The Crimean War in the 19th century did not only take part in the Black Sea, nor was it only about the Crimea.[29] The war had several stages and fronts. The Treaty of Paris, regulating the terms of peace after

24 D. Bönker, *Militarism in a Global Age: Naval Ambitions in Germany and the United States Before World War I* (2012); Berghahn, *Militarism: The History of an International Debate, 1861–1979* (1981).

25 C.W. Hughes, "Japan's Remilitarization and Constitutional Revision," in P.N. Stearns, *Demilitarization in the Contemporary World* (2013) 127–56.

26 Ibid., 6.

27 A. Bickford, "Demilitarization: Unraveling the Structures of Violence," in P.N. Stearns *Demilitarization in the Contemporary World* (2013) 23.

28 For a comparative result within the present project see: T. Koivurova and F. Holiencin, "Demilitarisation and Neutralisation of Svalbard: How Has the Svalbard Regime Been Able to Meet the Changing Security Realities During Almost 100 Years of Existence?," *Polar Record* 53, No. 2 (2017) 131–42.

29 R.L.V.F. Blake, *The Crimean War* (1971, reprinted 2006).

the Crimean War, was signed in 1856, by France, Great Britain and Russia. At that time, Finland had held the position of a Grand Duchy within the Russian Empire since 1809, after having formed part of the Kingdom of Sweden since the early Middle Ages. In an annexed convention to the Paris Peace Treaty, Russia undertook neither to build fortifications nor to create any military or naval establishments on the Åland Islands.[30] This convention and subsequent legal documents will be discussed in Chapter 2. While the demilitarisation regime was to some extent 'imposed' on both tsarist Russia (1856) as well as subsequently in Finland in 1921, it did not mean a factual occupation, or internationalised management of the region by international actors or other countries.[31] The relative unimportance of the Åland Islands was perhaps one of the immense advantages at the time. Russia needed to ensure its southern front and access to the Mediterranean while the Ottoman Empire was disintegrating. The fortress of Bomarsund on the north side of the main island of the Åland archipelago was apparently seen more as a boasting expression of presence and magnitude than an effective defence structure.[32] Sweden was eager to ensure that the islands were not used against the Stockholm capital area, but not eager enough to concur to the offer of the allies for sovereignty over the islands. Great Britain and France were ambivalent in their positions vis à vis Russia as well as the Ottoman Empire, but they wanted to minimise the risk of the Russian fleet reaching the Atlantic and their coasts.[33]

When Finland's sovereignty over the islands was confirmed by the League of Nations in 1921, provision was made for the territorial autonomy of the islands as well as for their demilitarisation and neutralisation. The

30 P. Wahlberg, ed., *International Treaties and Documents Concerning the Åland Islands 1856–1992* (1993).

31 C. Stahn, *The Law and Practice of International Territorial Administration: Versailles to Iraq and Beyond, Book* 57 (2010); R. Wilde, *International Territorial Administration: How Trusteeship and the Civilizing Mission Never Went Away* (2008).

32 G. Robins et al., *Bomarsund: Det Ryska Imperiets Utpost I Väster* (2004); A. Lambert, "Bomarsund i ett internationellt perspektiv. En fästningsuppgång och fall och dess plats i den globala konkurrensens strategiska kultur," *Åländsk odling 2004–2005*, 152–87.

33 The Danish Straits were, in fact, also regulated at the same time through two distinct treaties: *Treaty for the Abolition of the Sound Dues* between on the one hand Denmark, and on the other Austria, Belgium, France, Great Britain, Hanover, the Hansa Towns, Mecklenburg-Schwerin, the Netherlands, Oldenburg, Prussia, Russia and Sweden–Norway, signed in Copenhagen, 14 March 1857 (often referred to as the 1857 Copenhagen Treaty) and the separate *bilateral treaty between Denmark and the United States*, signed in Washington D.C., 11 April 1857. The treaties were discussed extensively in the case brought before the International Court of Justice, *Passage Through the Great Belt* (Finland v. Denmark) in 1991. P. Kleemola-Juntunen, "Straits in the Baltic Sea: What Passage Rights Apply?" (Forthcoming).

terms of this last aspect were to be specified in a separate international treaty, signed on 20 October 1921, which widened and reconfirmed the 1856 Convention.[34]

Before looking in more detail at these treaties in Chapter 2, the concepts utilised currently to describe security-related phenomena will be addressed.

1.3 The Birth and Impact of 'Comprehensive Security'

In recent years the conceptualisation of collective security and of war and peace have changed, as a result, *inter alia* of European integration, the 'end of the Cold War', the period of American hegemony, an expansive use of armed force in third countries, the war on terror, and the use of violence by Islamist and other extremist groups. It will be apparent in the following pages that new forms of security and civil-military co-operation and activities have been growing over a long period of time. While their roots can be identified many decades earlier, their observable reflections became visible in the 1990s. Altogether, such complex processes led to the endorsement of the concept of 'comprehensive security'.

The concept of 'comprehensive security' featured academically already in the 1980s with reference to the importance of environmental considerations for global peace and security as well as to the conditions of countries such as Japan.[35] The ideal, according to one argument, was that of going "[b]eyond war".[36] In 1986, the UN General Assembly adopted a resolution by which it remained engaged in what was described as a "comprehensive system of international peace and security".[37] The General Assembly was then concerned "at the tense and dangerous situation in the world and the danger of continuing down the path of confrontation and the arms race towards the abyss of the nuclear self-destruction of mankind" and proposed renewed engagement in the role of the UN and

34 *Rapport de la Commission Internationale de Juristes* (5 September 1920) in *Ålandsfrågan Inför Nationernas Förbund: La Question des Îles d'Aland*, Aktstycken (1920–1921).

35 J-P. Lehmann, "Japan's Quest for Comprehensive Security: Defence, Diplomacy and Dependence," *International Affairs (Royal Institute of International Affairs 1944–)* (1983); R.H. Ullman, "Redefining Security," *International Security* 8, No. 1 (1983) 129–53; D.S. Zagoria, "Beyond War: Japan's Concept of Comprehensive National Security," *Foreign Affairs* 64, No. 1 (1985) 193; A.H. Westing, "The Environmental Component of Comprehensive Security," *Security Dialogue* 20, No. 2 (1989) 129–34; T. Akaha, "Japan's Comprehensive Security Policy: A New East Asian Environment," *Asian Survey* 31, No. 4 (1991) 324–40; E.K. Stern, "Bringing the Environment In: The Case for Comprehensive Security," *Cooperation and Conflict* 30, No. 3 (1995) 211–37.

36 R.W. Barnett, *Beyond War: Japan's Concept of Comprehensive National Security* (1984).

37 UN GA Res. A/RES/41/92 (04.12.1986).

the peaceful settlement of disputes but also for the democratisation of international relations.

This effort was endorsed by the Soviet Union; in 1987, Vladimir Petrovsky, then deputy head of the Soviet delegation to the UN, forwarded an article written by Mikhail Gorbachev to the General Assembly titled "Reality and safeguards for a secure world".[38] It addressed issues of disarmament, confidence-building measures, human rights, economic co-operation and ended with mention of a need to bring a "system of comprehensive security into being". A few years later, in 1990, Petrovsky, then deputy foreign minister of the USSR, added to this extensive outline also an *aide-mémoire* in which he systematised Soviet views on possible ways for strengthening the UN and ensuring the primacy of international law.[39]

Since 1994, the United Nations Development Programme has used the concept of 'human security' in its annual reports as a more 'people-centred' concept. So, in addition to environmental concerns, the needs of specific countries, needs for democratisation and development, the idea of reinvigorating the United Nations and its work, were all aspects which informed this usage of the concept of 'comprehensive security' in the 1980s and 1990s. Nordic authors, in particular the 'Copenhagen School' in the field of International Relations, contributed to this shift.[40]

In international law, the new approach meant, first of all, the vast broadening of the possibilities to use force, with or without a UN mandate, for the purpose of a constantly enlarged understanding of international peace and security.[41] Peace activities and 'complex emergencies' involved civil-military interaction (CIMIC), which was approached in a rather managerial way as meaning effective protection and co-operation between civil and military actors, without dealing with the difficult issues of authority, legality, sovereignty, the self-determination of the regions and peoples affected and the sets of rules involved.[42] Furthermore, the comprehensive security

38 UN GA Res. A/42/574 (18.09.1987).

39 V. Petrovsky, "Towards Comprehensive Security Through the Enhancement of the Role of the United Nations (Aide-Mémoire)," *Alternatives: Global, Local, Political* 15, No. 2 (1990) 241–5.

40 O. Waever's co-operation with Barry Buzan is one such example. See e.g. his contribution in B. Buzan, *The European Security Order Recast: Scenarios for the Post-Cold War Era* (1990). Also several Nordic authors in W. Bauwens et al. (eds.), *Small States and the Security Challenge in the New Europe* (1996).

41 S. Spiliopoulou Åkermark, "The Puzzle of Collective Self-Defence: Dangerous Fragmentation or a Window of Opportunity? An Analysis With Finland and the Åland Islands as a Case Study," *Journal of Conflict and Security Law* 22, No. 2 (2017) 249–74.

42 OCHA, "Civil-Military Guidelines & Reference for Complex Emergencies," *United Nations* (2008); G. Hoogensen Gjørv, *Understanding Civil-Military Interaction* (2014).

approach allowed for an increased use of military means for meeting non-military threats, such as drug trafficking, uncontrolled migrations, environmental disasters and other 'non-traditional threats'. Terrorism, cyber security and hybrid warfare are often bundled together as more recent additions to this long list. One main consequence of this trend is, in the first place, increased uncertainty about when there is, legally, a status of war and, secondly, the blurring of the borderlines between what is military and what is civilian in the conduct of warfare, where for instance the same aircraft can be used for bombing one day and for the delivery of food and medicines the next.

The overall effect has been the fusion between war and peace and the prevalence of a permanent status of war reminiscent of an Orwellian state of affairs which permeates to varying degrees different parts of the world.[43] Such a state of affairs allows for revamped competition between alliances and superpowers. In addition, it allows for a constant state of exception which justifies deviations from international as well as constitutional rules.[44] There is renewed emphasis on military spending, military co-operation and the development of weapon technology.[45] Therefore, the idea of comprehensive security and its institutional implementations seems in the end to be undermining the goals it wanted to promote, most notably the respect of international legal rules and restrictions in the use of force.

Effects also became visible at the domestic level. During the Second World War, the Finnish army used various civilian assets in the conduct of war through confiscation or temporary usage, including those originating on Åland, from horses to cars and accommodation services for military personnel.[46] The point of departure in this 'traditional' case is that civilian assets could be used for military purposes in a state of exception. Today, there seems to be a reversal tendency of this basic assumption, in such a way that military resources and actors should be used to address non-military threats, in a situation which is reminiscent of a permanent state of exception. States of exception entail the possibility to deviate from rules protecting both the constitutional, that is, political, decision-making order as well as restrictions of individual rights.[47] One example can be that of environmental accidents,

43 G. Orwell, *Nineteen Eighty-Four: A Novel* (1949).

44 For one reaction see "A Plea Against the Abusive Invocation of Self-Defence as a Response to Terrorism," *Centre de Droit International, Université Libre de Bruxelles*.

45 I. Prezelj, "Comprehensive Security and Some Implemental Limits," *Information & Security* 33, No. 1 (2015) 13–34.

46 An example of this is the "Letter of 4 September 1939 From the Ministry of Interior to the Governor on Åland Concerning the Implementation of Relevant Legislation," *Länsstyrelsens arkiv*, Ea42a, Åland Archives.

47 D. Dyzenhaus, *The Constitution of Law: Legality in a Time of Emergency* (2006).

such as a major oil spill in waters around the Åland Islands. Such an accident is in principle to be countered with civil resources, the ministries of the Environment, Justice and the Finnish Environment Institute (SYKE). In an acute situation where no investments or planning are made on the civil side of environmental control, including management and surveillance, there shall not be any other avenue available than the use of military resources.[48]

While the earlier idea of *collective* security broadened the idea of security and conceptualised it as a common concern involving all member states of the UN, the idea of a *comprehensive* security broadened both the substantive areas where such collective security could be implemented as well as the actors that could and should be involved. The result of this process further leads to 'international society' taking over the domestic processes on the ground thereby strengthening perceptions of a democratic deficit.

These present-day trends are very different from the way international society acted in the case of the Åland Islands towards both tsarist Russia in 1856 as well as vis à vis Finland in 1921. One main objective of the League of Nations era was in fact to ensure that the Republic of Finland was as strong and sovereign as possible as a buffer towards the volatile and violent developments in revolutionary Russia. On none of these two occasions was there any need or interest in intruding excessively in the sovereignty of Finland (earlier of Russia) over the Åland Islands, even though some would argue that the creation of a territorial autonomy coupled with the demilitarisation and subsequent neutralisation of Åland were harsh enough conditions for the newly proclaimed Republic of Finland.[49] One could say that the 1940 bilateral treaty between Finland and the Soviet Union also confirms this general conclusion of a moderate attitude of international society and neighbouring countries based on restraint in the use of force. Firm but discrete Soviet presence and oversight of Finland's demilitarisation obligations through the establishment of a consulate were deemed strong enough, also for the powerful neighbour who had just won the Winter War. Such moderate approach which upholds state sovereignty deviates considerably from today's complex situations for instance in Kosovo, Crimea, Iraqi Kurdistan and Syria to mention but a few possible examples. Apart from the unique relationship between Finland and Åland, Sweden and Finland as well as Russia and Finland, all responses by actors around Finland and the Åland Islands hold evidence of a restraint in intrusion and in the use of force by which the Åland case is handled. Possible exceptions to this fact,

48 This scenario is discussed in the "Final Report of the Committee for a Revision of the Self-Government Act for Åland," Finnish Ministry of Justice, Report 33(2017) "*Ålands självstyrelse i utveckling*," 38.

49 Söderhjelm (1928).

including the Swedish expedition of 1918,[50] shortly after Finland's proclamation of independence, the dangerous and uncertain situations involved in both Russian revolutions and the manifold outcomes of the First World War, were met by a restoration of the demilitarised status as shall be seen in more detail in Chapter 2.

1.4 From the Cold War to a Multipolar World?

In addition to analysing the context in which the demilitarisation agreements were drafted, it is important to look at the contemporary situation and regional integration processes. After the Cold War, the situation in the Baltic Sea region changed radically. For example, the dissolution of the Soviet Union resulted in the new Russian Federation having limited access to the Baltic Sea. Former Soviet republics and Warsaw Pact members applied for membership in the EU and the North Atlantic Treaty Organization (NATO). There are currently three countries around the Baltic Sea which have not become members of NATO, namely Finland, Russia and Sweden. Norway, in contrast, is a member of NATO, but not a member of the EU.

After the Cold War, the world was typically understood as unipolar, with the United States as the sole superpower. The situation has, however, changed, with the rise of new powers such as the BRIC countries (Brazil, Russia, India and China). The United States may still be the most powerful country in economic and military terms, but it cannot make decisions without taking into account other important global factors and forces. The world today could even be described as multipolar, in contrast to how it looked during the Cold War; at that time, the United States and the Soviet Union were considered as forming two blocs in opposition, and other states had the choice either to join one of them or stay neutral. It was then that the non-aligned movement was created and the Finnish neutrality policy was seen as necessary. In the current multipolar world, it is not so self-evident if there are indeed any proper and meaningful sides to be taken. Many of the previously neutral countries have started to define themselves rather as 'non-aligned', implying non-membership in a military alliance. The foreign policy stance has been modified to include a declaration, for example that a certain state is not a member of NATO. Nevertheless, this does not necessarily seem to imply that such countries would abstain from other defence arrangements, for instance those in the EU, with neighbouring states or bilaterally (see Chapter 4).

50 A. Bondestam, *Åland Vintern 1918, Skrifter utgivna av Ålands kulturstiftelse*, Vol. 6 (1972).

Although neutralised, the Åland Islands are not neutral in the sense that they, or Finland, would abstain from taking sides in a war, nor that these islands have the rights and duties of neutral states under the international law of neutrality and humanitarian law.[51] However, the status of the Åland Islands is neutralised during war time, which means that no military action may take place in the region. It should be noted that the most important agreements regulating the use of the Ålandic territory originate from a period preceding the rise of the Soviet Union and the bipolar system. In fact, the Ålandic regime could be described as stemming from multipolarity: the 1856 agreement was between France, Great Britain and Russia in the context of the wider Paris Peace Treaty. The 1921 Convention in the League of Nations had Baltic Sea coastal states and European superpowers of the time as signatories.

In addition to the world becoming more multipolar, regional co-operation intensified after the Cold War, mainly in the non-military field. The EU, successor of the European Communities, has become a major economic and normative power, but co-operation also exists in smaller spheres. The Nordic countries have enlarged their co-operation in the field of defence, and the Council of the Baltic Sea States (CBSS) is one of the Baltic Sea regional organisations in existence since 1992. It includes the five Nordic countries, the three Baltic States, Russia, Poland and Germany. The CBSS focuses on soft security issues such as tackling cross-border crime, perhaps in the expectation of alleviating hard security threats. Military security is also excluded from the European Union Strategy for the Baltic Sea Region (EUSBR) as well as from the Northern Dimension, in the framework of which the EU co-operates with Russia, Iceland and Norway (see section 4.2). Despite the co-operation being mainly non-military, militarising tendencies have been witnessed in the region. The following section briefly introduces such trends and discourses of militarisation, while various forms of regional security co-operation are further discussed in Chapter 4.

1.5 Militarisation Trends and Discourses Today

With contemporary media debates on the deteriorated stability of the Baltic Sea, even the Åland Islands have been the subject of increasing attention in the news and politics in the neighbouring region.[52] Regional concerns are reflected in news about Russia deploying military equipment in Kaliningrad, Sweden remilitarising the islands of Gotland, increased NATO presence in

51 M. Gavouneli, "Neutrality—A Survivor?," *European Journal of International Law* 23, No. 1 (2012) 267–73.

52 K. Pynnöniemi and C. Salonius-Pasternak, "Security in the Baltic Sea Region: Activation of Risk Potential," *FIIA Briefing Paper No. 196* (2016).

the region and large military exercises taking place in the area. Such reporting reflects trends of both militarisation and securitisation, where increased military force is justified with alleged threats.

After the Cold War, the Baltic Sea did not attract much military attention, at least openly. The presence in the region of, for example, Russian as well as NATO military vessels and aircraft has increased in the past years, as evidenced by large military exercises by both sides. This may be one example of militarisation, a concept discussed earlier in this introduction. 'Militarisation' occurs, for example, when military presence is introduced to a geographical area which has been non-military for some time, such as the island of Gotland. In the technical sense, one may also talk about increased militarisation if a pre-existing military aspect intensifies. In scholarly literature, militarisation has occasionally been reduced to six features which reflect a technical understanding of militarisation: capability-based threat assessment, sensitivity to windows of opportunity, a tendency of rapid escalation of the use of force, an inclination to define territory in terms of the capture of defensible territory, marginalisation of non-military agencies (e.g., foreign affairs) and the predisposition to define strategy around the operating characteristics of available weapons.[53] Although militarisation thus formulated appears to be a relatively neutral, analytical or descriptive term, in politics it is used in a pejorative way by its critics. Officially, politicians deny that militarisation is consistent with national or European values.

Discourses on militarisation are particularly interesting in Finland, with a demilitarised province and a Cold War tradition of neutrality policy. Finland reformulated its neutrality policy as 'military non-alliance', and later 'non-membership of a military alliance'. These changes can mainly be attributed to Finnish accession to the EU in 1995; Finland has at times been active in furthering EU defence co-operation, at other times trying to prevent defence commitments (more on this in Chapter 4). Militarily non-allied Sweden and Finland facilitated the establishment of a European Security and Defence Policy (ESDP) by proposing in 1996 the move of the so-called Petersberg tasks from the Western European Union (WEU) to the EU. Established in 1999 through the Amsterdam Treaty, the ESDP originally only dealt with crisis management.[54] More substantial defence commitments were included in the Lisbon Treaty, adopted in 2007 (in force since 2009). To review the empirical discourse on militarisation and its reference points in Finland,

53 J. Schofield, *Militarization and War* (2007) 15.

54 T. Palosaari, *The Art of Adaptation: A Study on the Europeanization of Finland's Foreign and Security Policy* (2011) 12. The multiple rationales behind the Finnish and Swedish proposals and the ensuing developments in the period 1996–1999, leading to the wording adopted in the Amsterdam treaty, are beyond the scope of the present study.

the authors conducted a review of the Finnish parliamentary debates and their references to militarisation. In Finland, the left-wing parties have been particularly active in criticising the militarisation of EU structures. For example, a Left Alliance MP stated in 2001 that the "third critical part in EU development is that the EU is aimed to be militarised".[55] It appears that militarisation constitutes something negative for the critics in the case of an issue that was previously not a military one. By the same token, the Finnish Foreign Minister at the time, Erkki Tuomioja (Social Democrat), stated in a parliamentary debate in December 2014 that Finland did not want to participate in militarising civilian activities.[56]

Even the parties that supported further military measures attached a negative association to the concept of militarisation and denied that military activities constituted militarisation. The comments of the ruling Centre Party regarding militarisation mainly stated that a decision made by the government does not imply militarisation or anything of the kind. For example, a Centre Party MP argued that using the Defence Forces in executive assistance missions is "common sense thinking" instead of "anything pointing towards militarisation".[57] The Defence Minister of the same party also stated that increasing the budget of the Finnish Defence Forces is not "militarisation or anything like that".[58] According to the same party representative, providing attitudinal training for minors on defence issues does not refer to "any militarisation or related activities".[59]

It is worth noting that the number of militarisation references is not very substantial, and references to the militarisation of the EU were absent before the turn of the millennium. After that, militarisation references related mainly to the 'mutual assistance' clause that was being prepared and later included in the Lisbon Treaty. Some politicians seemed to be concerned over civilian crisis management becoming militarised in the EU context. Militarisation was thus seen as an alleged aim, which either European or Finnish politicians apparently sought to enhance.

Although militarisation as a concept is thus not very much utilised in Finnish political debates, tendencies pointing towards militarisation are visible in

55 Finnish Parliament, "Minutes of the Plenary Session on 10 December 2001," 10 December 2001.

56 Finnish Parliament, "Minutes of the Plenary Session on 11 December 2014," 11 December 2014.

57 Finnish Parliament, "Minutes of the Plenary Session on 22 November 2006," 22 November 2006.

58 Finnish Parliament, "Minutes of the Plenary Session on 14 Sepember 2006," 14 September 2006.

59 Finnish Parliament, "Minutes of the Plenary Session on 10 October 2006," 10 October 2006.

Table 1.1 Militarisation in the Finnish parliamentary debates 1995–2016.

Party	*Reference to militarisation*
Social Democrats	EU (2), foreign policy (2), peace-keeping (1), society (2), civilian activities (1)
Green Party	EU structures (1)
Left Alliance	EU (17), security in the EU (1), civilian crisis management (2), societal atmosphere (1), Finland (1), society (2), non-military service (1)
Christian Democrats	EU's fundamental nature (1)
Centre Party	Executive assistance (1), defence resources (1), defence training for minors (1), Arctic (1)

other regional co-operation outside of the EU (these are further discussed in Chapter 4). Although interest in regional defence co-operation clearly exists, the two strongest supporters of the demilitarisation of the Åland Islands, Sweden and Finland, seem to have decided to stay outside NATO. NATO membership would appear, however, to be a contentious and much-debated issue, *inter alia* in Sweden and Finland, but there is no majority support for membership at the moment: approximately a fourth of Finns and a third of Swedes support NATO membership.[60] Both countries co-operate though with NATO under the Partnership for Peace programme and they have also hosted major military exercises with NATO members' participation in recent years.

1.6 Geopolitics and Securitisation as Potential Justifications for Militarisation

In the previous section, we discussed some potential manifestations of militarisation, but how do such processes come about? When reviewing scholarly literature dealing with increased emphasis on military affairs, two concepts crop up particularly often: geopolitics and securitisation. These two concepts relate to militarisation, but encompass more than just military security; they reflect the ongoing processes in international relations, but from two different perspectives. In other words, scholars focusing on what is done in politics tend to focus more on geopolitics, while those focusing on what is said tend to emphasise securitisation trends. Securitisation, geopolitics and militarisation are thus conceptually connected, although not necessarily discussed together. Whereas militarisation often refers more to

60 H. Vuorela, "Nato-jäsenyyden vastustus väheni—Kanerva yllättyi muutoksen vähäisyydestä," *Maaseudun Tulevaisuus* (11 January 2017); J. Gummesson, "SvD/Sifo: Kraftigt ökat motstånd mot Nato," *Svenska Dagbladet*, 2 July 2016.

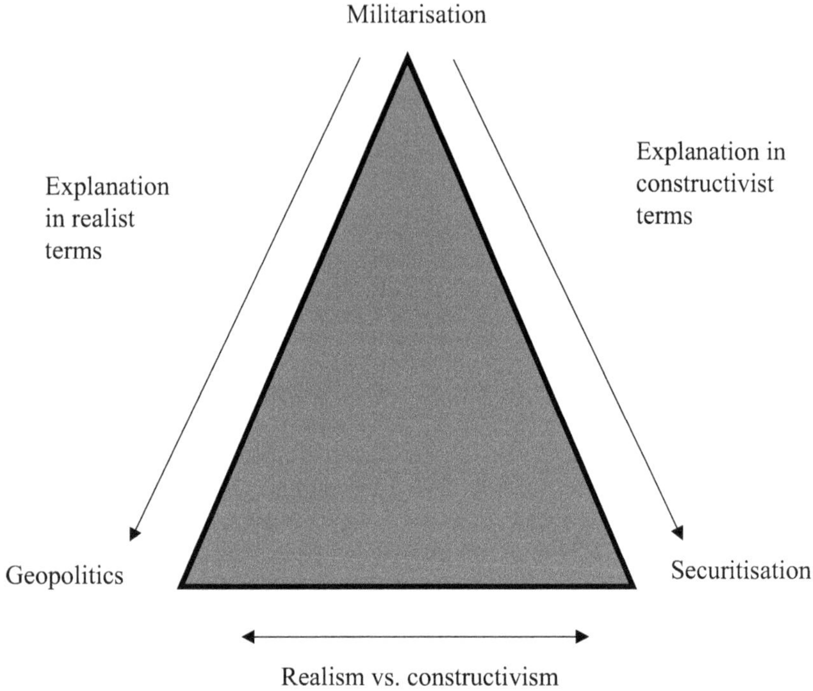

Figure 1.1 International relations view on militarisation explained by geopolitical and securitisation perspectives.

the adoption of practical military measures, geopolitics and securitising justifications, *inter alia* can serve as different approaches to look at militarisation, one from a realist and the other from a constructivist perspective.

Constructivist international relations scholars tend to discuss securitisation to describe the process of justifying more military emphasis with threat discourses. Securitisation thus denotes a process with which exceptional measures are sought to be justified with existential threats, originally developed in the so-called Copenhagen School of International Relations.[61] Constructivism, simply put, refers to the school according to which reality is socially constructed and in which security "is the discursive practice of framing an issue as a security issue".[62] Desecuritisation, in turn, is a process which "restores the possibility of exposing the issue to the normal haggling and

61 B. Buzan et al., *Security: A New Framework for Analysis* (1998).

62 O. Read, "How the 2010 Attack on Google Changed the US Government's Threat Perception of Economic Cyber Espionage," in *Cyberspace and International Relations: Theory, Prospects and Challenges*, eds. J-F. Kremer and B. Müller (2014) 208.

questioning of politicization".[63] Securitisation can be conducted, *inter alia*, by the media and by politicians who justify certain exceptional measures on the basis of threats.[64] It need not be manifested through military action, but it is often the case that the military is presented as the most effective solution to tackle threats. For example, in arguing for increased military co-operation in the EU, the war in Ukraine and the annexation of Crimea have been utilised as threatening examples.[65] With regard to the demilitarised Åland Islands, securitisation could be observed, for example in the comments of the Finnish Defence Minister of "green men" potentially entering the Åland Islands.[66] In the constructivist school, the emphasis is often on discourses, but securitisation can also be looked at from a sociological perspective, that is, whether such securitisation has been successful.[67]

Whereas a widespread consensus of securitisation as the "process of making an issue a 'security' issue" can be observed,[68] geopolitics is utilised in much more controversial ways. It can simply be denoted as a form of geography and politics in the context of social reproduction of power and political economy.[69] If international relations are approached from a geopolitical perspective, the ontological premises are usually realist, that is, interstate relations are seen to consist of great power rivalry.[70] A simple definition for geopolitics refers to the "distribution of power in physical space", but that does not say anything about where the concept is used.[71] In actual discourse, geopolitics often implies the possibility of using military power against a potential aggressor,[72] but it has also been connected to the mere impact of geographical factors on states or the

63 O. Waever, "The EU as a Security Actor: Reflections From a Pessimistic Constructivist on Post-Sovereign Security Orders," in *International Relations Theory and the Politics of European Integration: Power, Security and Community*, eds. M. Kelstrup and M.C. Williams (2000) 224.

64 J.A. Vuori, *How to Do Security With Words: A Grammar of Securitisation in the People's Republic of China*, *Turun Yliopiston Julkaisuja* (2011).

65 S. Heinikoski, "'Pool It or Lose It'—Discourses on EU Military Integration and Demilitarisation in the Baltic Sea," *Journal on Baltic Security* 3, No. 1 (2017) 32–47.

66 M. Moberg et al., "Vaara! Vihreitä miehiä Ahvenanmaalla!—Näin se voisi tapahtua," *Suomen Kuvalehti*, 2 August 2015.

67 e.g. H. Carrapico, "Analysing the European Union's Responses to Organized Crime Through Different Securitization Lenses," *European Security* 23, No. 4 (2014) 601–17.

68 A.J. Bellamy, "Security Communities and Their Neighbours: Regional Fortresses or Global Integrators?" (2004) 30.

69 G.Ó. Tuathail and S. Dalby, "Introduction: Rethinking Geopolitics—Towards a Critical Geopolitics," *Rethinking Geopolitics* (1998) 2.

70 S. Dalby, "Realism and Geopolitics," in *The Ashgate Research Companion to Critical Geopolitics*, eds. Klaus Dodds et al. (2013) 33–47.

71 L. Simón, *Geopolitical Change, Grand Strategy and European Security* (2013) 33.

72 M. Wigell and A. Vihma, "Geopolitics Versus Geoeconomics: The Case of Russia's Geostrategy and Its Effects on the EU," *International Affairs* 92, No. 3 (2016) 605–27.

relation between geography and international relations.[73] If reference is made to geopolitics returning to the global agenda, it is usually in the context of power politics and military balance. Some have argued that after the Cold War, geopolitics was no longer relevant, but the prevalence of geopolitics in recent debates would suggest the contrary.[74]

Geopolitical thinking implies that the world is understood as being comprised of competing powers with a focus on hard security, that is, use of the military. Geopolitics can manifest itself in increased military preparation being called upon in order to correct power imbalances. The demilitarisation and neutralisation of the Åland Islands is in contrast to this tendency, although it can be simultaneously argued to originate from geopolitics. The demilitarisation of the Åland Islands excludes all military presence on the islands, but Finland has the right to defend the islands also militarily if they are attacked. International agreements intend to prevent any geopolitical threats to the islands, and it cannot be said that such threats materialised even during the two world wars.

In the 1850s after the Crimean War, the great powers Russia, Great Britain and France decided to demilitarise the Åland Islands. It appeared indeed to be a question of geopolitics: as a result of the islands' geographical importance at the time, they were demilitarised in order to prevent further conflicts. Russia was forced by Great Britain and France to demilitarise the islands. The agreement can be explained through geopolitical and realist terms: all actors had an interest not to put resources in the Baltics, which made demilitarisation a rational solution. As was the case of the Åland Islands, demilitarisation can thus be seen to function as a geopolitical instrument. To some extent, demilitarisation is also a reminder of the prevalence of geopolitics in some political circles and in some periods of time. The competing interests in a region may be so intense that banning the very presence of military force is the only option. Although the intention of this book is not to provide an in-depth analysis of the processes of militarisation, geopolitics and securitisation, these processes provide a useful conceptual background particularly for discussing regional security co-operation and the Åland Islands in Chapter 4. Before that, the legal framework of the demilitarisation and neutralisation regime will be discussed in Chapters 2 and 3.

73 L. Simón, *Geopolitical Change, Grand Strategy and European Security* (2013).
74 Tuathail and Dalby, "Introduction: Rethinking Geopolitics—Towards a Critical Geopolitics" (1998).

2 The Legal Regulation of the Demilitarisation and Neutralisation of the Åland Islands

The purpose of the present chapter is to explore aspects of continuity and change in the Åland regime of demilitarisation and neutralisation, as well as its contemporary status, rather than to explore in detail the material scope of the demilitarisation and neutralisation regulations. This material scope of the rules is also infused in Chapters 4 and 5 of the book. The present chapter does not offer a systematic analysis of all the specific rules in the various conventions, but rather gives an account of their core essence, evolution and dynamics in the past and today.

2.1 An Overview of the International Agreements Establishing the Demilitarisation and Neutralisation of the Åland Islands

The demilitarisation and neutralisation of the Åland Islands is a regime involving a series of regulations, institutions and processes at various levels, national as well as international. This regime also entails the involvement of several actors and states parties; it is a nexus of interlocked regimes. This often comes as a surprise to many observers who think in binary, mutually exclusive patterns. In a binary way of thinking, you can only have one regime or one set of legal rules applying at one time. This is, however, not the case in practice. When the term 'regime' is used in the current work, the assumption is not that international affairs are a progressive co-operative process, exempt from power relations and hegemonic tendencies or behaviour.[1] On the contrary, the complex regime applying to the Åland Islands

1 We note in this regard the gap between North American literature and European interpretations in international law and relations. See also Koskenniemi's critique in the final chapter of M. Koskenniemi, *The Gentle Civilizer of Nations: The Rise and Fall of International Law 1870–1960* (2001); M. Noortmann, *Enforcing International Law: From Self-Help to Self-Contained Regimes* (2016) 132–46, originally pubished in 2005.

is a paradigmatic example of multilevel governance where local (Ålandic), national, international and European norms and institutions co-exist, co-operate, collide and compete and where legal, political, diplomatic and military realities run parallel.

Åland's demilitarisation and neutralisation is made up of a series of documents that are all still considered to be valid and in force.[2] The 1856 Convention on the demilitarisation of the Åland Islands was attached as an appendix to the Paris Peace Treaty at the end of the Crimean War, and included three states parties, namely France, Great Britain (including at that time Ireland) and Russia. It consists of one single operative article for the purpose of "general peace" *(les bienfaits de la paix générale)* in which the parties agree not to fortify the islands and not to retain or create any establishment of a military or naval nature (Article 1).[3]

This obligation has been referred to as the 'Åland servitude' with the effect of underscoring that the demilitarisation regime is connected to the territory of Åland and not to any individual state as the responsible party for demilitarisation.[4] However, since this terminology is not used in the convention itself, the authors avoid making reference to it. The original parties to the convention were Russia—which Finland (including the Åland Islands) was part of until 1917—together with Great Britain and France who were at the time co-operating within the framework of the Crimean War(s). Neither Sweden, due to its policy of neutrality and cautious attitude in the struggles between the superpowers, nor Finland (which was yet to be an independent state) was party to the 1856 Convention. Söderhjelm argued that the convention was binding not only for the three signatory powers, but also for the other parties to the 1856 Paris Peace Treaty to which the

2 M. Björkholm and A. Rosas, *Ålandsöarnas demilitarisering och neutralisering* (1990); L. Hannikainen, "The Continued Validity of the Demilitarised and Neutralised Status of the Åland Islands," *Zeitschrift Für Ausländisches Öffentliches Recht Und Völkerrecht* 54, No. 3 (1994) 614–51; U. Linderfalk, "International Legal Hierarchy Revisited—The Status of Obligations *Erga Omnes*," *Nordic Journal of International Law* 80, No. 1 (2011) 1–23. In addition to the 1856 convention, the core agreements discussed are: "Convention Relating to the Non-Fortification and Neutralisation of the Åland Islands (1921)," Finnish Treaty Series (FTS) 1/1922: English translation available in 17 AJIL 1923, Supplement: Official Documents, 1–6; "Treaty Concerning the Åland Islands Between Finland and the Union of Soviet Socialist Republics," FTS 24/1940; "Reactivation of the Previous Treaty Between Finland and the Union of Soviet Socialist Republics," FTS 9/1948; "Peace Treaty with Finland," FTS 20/1947: English translation available in 42 AJIL 1948, Supplement: Official Documents, 203–23. Most of these documents can also be found electronically at the website of the Åland Parliament.

3 In the French original: "*[. . .] les îles d'Aland ne seront pas fortifiées, et qu'il n'y sera maintenu ni créé aucun établissement militaire ou naval.*"

4 Björkholm and Rosas (1990) 26–7.

demilitarisation convention was appended, that is, for Austria, Prussia and Sardinia.[5]

The Commission of Jurists, which gave a consultative view to the Council of the League of Nations on the so-called Åland Question (see section 2.2), took the view that the 1856 Convention was still in force in 1920, that it had established rules the respect of which all interested states (and not only the signatories) could claim, while any state holding sovereignty over the islands would be bound by this "demilitarisation system *(le système de demilitarisation)*".[6] Interestingly, the lawyers speak here of a 'system' even though the 1856 Convention only included one operative article and provided for no mechanism of monitoring or enforcement. However, it formed part of a larger context, that is, the Paris Peace Treaty terminating the various segments of the Crimean War. The assessment may also have been anachronistically influenced by the knowledge of the number of agreements affecting the Åland Islands concluded during the various stages of the First World War.[7]

The 1856 Convention does not distinguish between war time and peace time. The prohibition of military or naval establishments covers both, but did not explicitly prohibit the presence of military forces or military activities. There is some evidence that from the outset, at least France and Sweden had a broader understanding of the obligations created by the 1856 Convention. Soon after the completion of the Paris Peace Treaty, Napoleon III had assured that it meant that "*Les Russes ne peuvent avoir aucun établissement militaire ou maritime là, ne peuvent pas non plus y réunir des troupes*".[8] This ambiguity was soon proved to be a limitation in the material scope of the convention and was addressed in 1921.

The Crimean War stood in the watershed between old-fashioned methods and weapons and modern warfare.[9] Among the many new inventions were rifled small-arms and artillery, aerodynamically designed ammunition, the electric telegraph, the railway, steam ships rather than sail-driven ones and iron-clad warships. This technical revolution changed the premises of

5 J.O. Söderhjelm, *Démilitarisation et neutralisation des Iles d'Aland en 1856 et 1921* (1928) 100–17.

6 Commission Internationale de Juristes, *Ålandsfrågan Inför Nationernas Förbund: La Question des Îles d'Aland* (1920) 278. The jurists use the term "*un règlement d'intérêts européens*". Max Huber became soon thereafter judge at the Permanent Court of International Justice and arbiter in the Palmas case. On Huber's pacifist agenda see J. Delbrück, "Max Huber's Sociological Approach to International Law Revisited," *European Journal of International Law* 18, No. 1 (2007) 97–113.

7 Söderhjelm (1928).

8 Cited in Söderhjelm (1928) 117. In translation: "The Russians cannot have any military or naval installation there, and they cannot gather any troops there either".

9 R.L.V.F. Blake, *The Crimean War* (1971) 145.

society as a whole, including for international law. There was a more evident and pressing need and goal to limit and control war and to create conditions and ethics of peace. The institutional tool of demilitarisation was a small step in this direction. It allowed for self-limitation by states parties and contractual, reciprocal agreement on limitations with respect to armament and fortification issues, shortly after von Mohl launched the theory of constitutional self-limitation of the state and of the primacy of law.[10] The institution of demilitarisation had the additional effect of moving and reconceptualising a potential hot spot from the military end of a spectrum of possible alternatives towards the political and diplomatic field in this pendulum between war and peace. The pendulum in the relations between the political, diplomatic and military had been already described by von Clausewitz.[11]

The 1921 Convention on the Non-Fortification and Neutralisation of the Åland Islands[12] was adopted as one of the outflows of complex processes that involved the declaration of independence of the Republic of Finland (1917) and the dispute between Finland and Sweden on sovereignty over the islands, a dispute which was eventually brought before the League of Nations subsequent to discussions at the Paris Peace Conference (1919–1921). The 1921 Convention had been recommended by both commissions appointed by the Council of the League of Nations (the Commission of Jurists and the Commission of Rapporteurs). The original parties to this treaty were (in alphabetical order) Denmark, Estonia, Finland, France, Germany, Great Britain (including Ireland), Italy, Latvia, Poland and Sweden. This convention is the most detailed of the international treaties concerning the demilitarisation and neutralisation of the Åland Islands, including nine operative articles (see section 2.2). It applies in peace time and in war time, this being one of the reasons why a new treaty was considered necessary, in addition to bringing Finland, Sweden and Germany as well as other states around the Baltic Sea into the agreement. Russia is not to be found among the parties of the 1921 Convention given that revolutionary Russia had not yet been recognised by most states, nor was it at the time a member of the League of Nations. As a result of its detailed content and wide range of states parties, it is understood as being the hub of the demilitarisation and neutralisation of the Åland Islands.

10 R. von Mohl, *Die Polizei-wissenschaft nach den Grundsätzen des Rechtsstaates*, 2nd edn. (1844).

11 K. von Clausewitz, *Vom Kriege* is included in his collected works published as *Hinterlassene Werke über Krieg und Kriegführung* (1832–1837). For a recent analysis see T. Waldman, *War, Clausewitz and the Trinity* (2013).

12 Also Söderhjelm (1928) 372–8.

The Treaty between Finland and the Soviet Union concerning the Åland Islands was concluded in October 1940 after Finland's defeat by the Soviet Union in the so-called Winter War (winter of 1939–1940).[13] It has three operative articles, including a right of the Soviet Union to establish a consulate on Åland; the consulate operates still today. Sweden has had a consul on the Åland Islands since 1871.[14] Not only is the presence of the Russian consulate evidence of the continued validity of the 1940 treaty, it was also explicitly confirmed via diplomatic note to Finland from the Soviet Union in 1948 as well as through the 1992 Protocol (signed in Helsinki on 11 July 1992) regarding the continued validity of selected bilateral agreements at the time of the dissolution of the Soviet Union and the succession by the Russian Federation. The importance of this treaty for the Soviet Union is also highlighted by the fact that it was registered by the Soviet Union by notification to the Treaty Office of the UN (13 March 1948) and was thereafter published in the United Nations Treaty Series in 1950 as a bilateral treaty.[15]

In the late stages of the Second World War and after the so-called Continuation War (1941–1944) between Finland and the Soviet Union, Finland found itself among the defeated powers, after co-operating with Nazi Germany. The Armistice Agreement of September 1944 between Finland on the one hand and the Soviet Union and the United Kingdom of Great Britain and Northern Ireland on the other, meant not only considerable reparations to be paid by Finland, but also the reconfirmation of the bilateral treaty of 1940 concerning the Åland Islands.[16] The final Treaty of Peace between Finland and the Allied and Associated Powers (in alphabetical order: Australia, Byelorussian Soviet Socialist Republic, Canada, Czechoslovakia, Great Britain, India, New Zealand, the Union of Soviet Socialist Republics, Ukrainian Soviet Socialist Republic, United Kingdom of Great Britain and Northern Ireland, Union of South Africa) was only signed in Paris in 1947.[17] This treaty provided in its Article 5 that the Åland Islands shall remain demilitarised "in accordance with the situation as at present existing", a formulation which allowed for the nexus of previous legal obligations with various states to remain intact and run in parallel. In addition, this

13 FTS 24/1940.

14 Nikolai Sittkoff, shipowner and successful businessman in Mariehamn, became the first Swedish consul on Åland. He represented the Union of Sweden and Norway until 1887. "*De svenska konsulaterna i Mariehamn*," unpublished (and undated) and "*Sveriges generalkonsulat i Mariehamn*" (dated 15.01.2014), two overviews by the Swedish Consulate in Mariehamn, on file with the authors.

15 67 UNTS (1950), No. 872, 139–51.

16 FTS 4/1944.

17 FTS 20/1947.

arrangement meant that Finland remained in a bilateral relationship towards the Soviet Union, while the older legal obligations (the 1856 and 1921 treaties) would continue to apply for states such as Great Britain or France.

By necessity, this brief overview of treaties constitutes a simplification of the multiple legal and political relations created by the various legal undertakings, not only on the basis of contractual legal interstate obligations but also on the basis of customary international law or even *jus cogens*. This has resulted in Finnish politics often referring overall to the demilitarisation of the Åland Islands 'under international law', as was done for instance in the process of the ratification of the Lisbon Treaty in the Finnish Parliament.[18] Bilateral, multilateral and customary rules co-exist. The Accession Treaty of the Republic of Finland to the European Union (1994) included in its Protocol No. 2 provisions pertaining to the Åland Islands and referring not simply to the autonomy of the islands, but to "the special status that the Aaland islands enjoy under international law". Similar references have been made on many occasions, some of which are mentioned throughout this book.

There were also numerous other points of time when agreements and international documents as well as national legislation have had a profound impact on the situation prevailing in the Åland Islands. One could mention here in particular the series of treaties ending the First World War in 1918 and prior to the outcome of the Versailles peace talks, namely the Treaty of Brest-Litovsk (3 March 1918) between Germany, Austria-Hungary, Bulgaria and the Ottoman Empire on the one hand and Russia on the other and shortly after, the Treaty of Peace between Finland and Germany of 7 March 1918.[19] A similar agreement was signed in December 1918 between Germany, Finland and Sweden with the exclusion of revolutionary Russia.[20] The arguments presented by various sides during the First World War on the temporary or limited nature of military fortifications and actions were insufficient to prohibit the escalation of such activities, despite the fact that the dynamics and justifications of the war itself have multiple explanatory levels: the neutral position of Sweden, the proclamation of the Republic of Finland in December 1917 followed by the civil war in 1918, together with the incremental disintegration of the Russian empire and the turbulent times of the Russian revolutions. While states parties have occasionally tried to

18 *Regeringens proposition till Riksdagen med förslag om godkännande av Lissabonfördraget* (RP 23/2008 rd, 116). See also the declaration made by the representative of Finland in the EU Committee of permanent representatives (COREPER) in December 2009, Council of the European Union, Summary Record, 16980/09, CRS CRP 45, dated 11 January 2010.

19 FTS 3/1918; Söderhjelm (1928) 133 and in appendix. See also the Treaty of Brest-Litovsk.

20 Söderhjelm (1928) 134.

maximise advantages for themselves, the core of all these agreements, and over a long period of time, has repeatedly confirmed that fortifications and military troops should not be found on the islands and the islands should not be used for military purposes.[21]

2.2 The 1921 Convention on the Non-Fortification and the Neutralisation of the Åland Islands

The original commitments by the contracting parties in the 1856 demilitarisation agreement and the Paris Peace Treaty were broadened by the Decision of the Council of the League of Nations of 24 June 1921 which in turn resulted in the 1921 Convention on the Non-Fortification and Neutralisation of the Åland Islands, negotiated under the auspices of the League of Nations.[22] Indeed, the Commission of Jurists as well as the Commission of Rapporteurs of the Council emphasised the need to negotiate a more encompassing treaty in addition to securing an autonomous status for the archipelago. The Commission of Jurists devoted considerable attention to the issue of demilitarisation, focusing on the one hand on the question of the validity of the 1856 demilitarisation convention, and on the other hand on the relevance of the 1856 undertakings under the circumstances prevailing at the time.[23] This is somewhat surprising in view of the fact that the core issue it was facing was the question of the possible jurisdiction of the League of Nations in the sovereignty dispute between Finland and Sweden. The fact that they devoted a section of their opinion on matters of demilitarisation highlights the importance they attributed to the issue.

Among the contextual factors noted by the Commission of Jurists in its examination was the recognition of Finland by Great Britain (which took place first in May 1919 and unconditionally on 21 January 1920), much later than the recognition of Finland by the Soviet Central Executive Committee in Petrograd on 4 January 1918. The Soviet Committee's stance had been based on the proclamation of the principle of self-determination of foreign peoples (*les peoples allogènes de la Russie*) of 15 November 1917. Soviet recognition was immediately followed by Swedish and French recognition of Finland that same day. The emanation of this new republic was

21 Söderhjelm (1928). On the military activities on Åland at the time see also M. Isaksson, *Ryska positionen Alandskaja* (1983); B. Stjernfelt, *Ålands hav hch öar—brygga eller barriär?: Svensk-Finsk Försvarsfråga 1915–1945*, Vol. 72 (1991); K. Gustavsson, *Ålandsöarna—en säkerhetsrisk? Spelet om den demilitariserade zonen 1919–1939* (2012) 18–50.

22 LNTS No 255, Vol. IX (1922) 211–22.

23 *Ålandsfrågan inför Nationernas förbund: La Question des Îles d'Aland*. Vol. I (1920) 265–79.

described by the Commission of Jurists as a "new political phenomenon". Importance was attached to the fact that the Russian Empire had not been able to revoke the 1856 Convention, despite efforts to do so in 1906, and that the demilitarisation had been reaffirmed in the so-called Baltic Declaration of April 1908 between Germany, Denmark, Russia and Sweden.[24] While not pronouncing on the exact scope of the 1856 agreement, the Commission of Jurists found that the presence of Russian troops and fortifications during the First World War was not permitted by the convention. While the goal of the 1856 Convention had not been to neutralise the islands in times of war—in contrast, for instance, to the position of the Ionian Islands in the Mediterranean which had been fully neutralised in 1863–1864—Russia had assured Sweden, Great Britain and France that the fortifications had a temporary character. The temporary nature was confirmed by their demolition following Article 6 of the Treaty of Brest-Litovsk, as well as the agreement of December 1918 between Sweden, Finland and Germany, noted the Commission of Jurists. The Commission of Jurists concluded that the provisions of 1856 had the character of norms of a European interest ("*un règlement d'intérêts européens*"), which could not be terminated or altered through unilateral acts or through acts involving only a few of the parties to the various agreements.[25]

The aim of the 1856 Convention was to eliminate the possibility that too heavy an influence ("*une influence trop grande*") is obtained by the state controlling the islands. Furthermore, the jurists looked at the 'special status' of the islands internationally and spoke of a 'particular statute'. As long as these norms are not replaced, each interested state has the right to claim that they should be respected. This indicates how the Commission of Jurists combines ideas of factual control and sovereignty, contractual obligations and general international law, and looks at power relations at the same time as upholding the need to prevent war through legal norms and institutions. This is, of course, of no surprise. The President of the Commission of Jurists, the French lawyer and professor of international law in Paris, Ferdinand Larnaude, had participated in the League Commission which met in 1919 to prepare the League of Nations charter. French Prime Minister Georges Clemenceau was one of the driving forces.[26] Larnaude was joined by Max Huber, advisor to the Swiss government and professor of international law.[27] Finally, professor of constitutional and international law in Amsterdam Antonius Struycken was a member of the Dutch Council of

24 Söderhjelm (1928) 118–23.

25 *Ålandsfrågan inför Nationernas förbund: La Question des Îles d'Aland.* Vol. I (1920) 275.

26 B. Stråth, *Europe's Utopias of Peace* (2016) 254–55.

27 Delbrück, "Max Huber's Sociological Approach to International Law Revisited," *European Journal of International Law* 18(1) (2007).

State and an active voice in the establishment of the League of Nations after the great disillusion of the First World War.

The Commission of Rapporteurs, which took over once the matter of international jurisdiction was settled, also chose to dedicate part of their opinion specifically to issues of demilitarisation. The commission was comprised of Belgian Minister and diplomat Baron Beyens,[28] Felix Calonder, Swiss lawyer and politician and Abram Elkus, Jewish-American lawyer. They argued for the negotiation and adoption of a convention dedicated to the demilitarisation and neutralisation of the Åland Islands and had in front of them drafts proposed by Finland and Sweden respectively.[29] The commission did not wish to pronounce on what they saw as technical military and naval matters but believed that, contrary to the original views of Finland, policing presence would suffice to maintain security locally. As they put it, "*moins il y aura d'appareil militaire à Aland, plus la tranquillité y sera assurée*"—the less military presence there is on Åland, the more guarantee of calm on the islands.

The main negotiations for the new demilitarisation treaty were held in October 1921 in Geneva.[30] The decision of the Council of the League of Nations (24 June 1921) had referred explicitly to the particular interests of Sweden as well as to the first draft treaty presented by Sweden. The demilitarisation conference was opened by the Chinese President of the Council of the League of Nations, Vi Kyuin Wellington Koo. In his opening speech, he acknowledged that "the neutralisation and disarmament correspond fully to the goals of the League of Nations, i.e. the development and safeguarding of international peace".[31] The Danish diplomat Herman Anker Bernhoft was thereafter elected as chair of the conference; he had earlier been involved in the negotiations concerning North Schleswig/Sønderjylland within the framework of the Treaty of Versailles, and later headed the Danish delegation in the East Greenland Case before the Permanent Court of International Justice (1932–1933). The Secretariat of the conference was headed by Donisio Anzilotti.

Revolutionary Russia was not among the states negotiating in Geneva. In January 1918, the Russian Socialist Federative Soviet Republic had been proclaimed in Moscow, as a predecessor to the Union of Soviet Socialist Republics (USSR), but Russia had plunged into what has been called a "concatenation of civil wars".[32] While the new state had been recognised

28 E.-N. Beyens baron, *L'Allemagne avant la guerre* (1915).

29 *Ålandsfrågan inför Nationernas förbund: La Question des Îles d'Aland.* Vol. II (1920) 127–93.

30 Société des Nations, "Conférence Relative à la Non-Fortification et à la Neutralisation des Iles d'Aland," *Actes de la conférence* (1921) C.554 VII.

31 Translation by the authors.

32 M.A. Reynolds, *Shattering Empires: The Clash and Collapse of the Ottoman and Russian Empires 1908–1918* (2011) 167–90.

by Finland and Estonia in 1920 through the Treaty of Dorpat, most other countries and members of the League of Nations had not yet done so.[33] The USSR was, in fact, to only become a member of the League of Nations much later, during 1934–1939.

In addition to the diplomats involved in the negotiations in Geneva, naval experts who worked with the geographical delimitation of the Åland Islands zone were also present. Each delegation comprised both diplomatic as well as military delegates. One of the early issues in the conference debates was the proposal by French delegate Jean Gout that the delimitation should be carried out with the assistance of a technical commission, which prompted the immediate reaction of General Oscar Paul Enckell, head of the Finnish delegation to the conference. The Swedish representative, Eric Birger de Trolle, when introducing the Swedish proposals, highlighted what he saw as the main difference in the original proposals by his country and those of Finland; namely, for Finland, the difference was in the approach towards their rights to maintain troops on the islands and for Sweden, such a right would not be acceptable since it went far beyond the norms established by the 1856 Convention as well as against the ideas of the Commission of Rapporteurs described above.[34] The Swedish representative drew attention to the fact that the inhabitants of the islands were exempted from military service under the Act of Autonomy which had been adopted in 1920 in Finland.[35] Such military presence, he argued, would upset the Ålanders. The Polish delegate, Syzmon Askenazy, proposed in the second session (11 October 1921) that representatives of states non-riparian in the Baltic Sea should attempt to seek a compromise between Finnish and Swedish positions. The German representative Oskar Trautmann insisted that the interests of Germany were not to be forgotten. The Polish proposal for a small preparatory commission was endorsed.[36] A convention draft was presented by this commission on 14 October 1921 prompting discussion on the geographical co-ordinates and on the choice between a three or, alternatively, four nautical miles' rule for the scope of the demilitarised and neutralised zone (see Chapter 3). The Italian representative, Arturo Ricci-Busatti, was eager to retain the right to 'inoffensive passage', but it was clarified in the discussions that submarines exercising such a right should be navigating on the surface.

33 E.H. Carr, *A History of Soviet Russia: The Bolshevik Revolution, 1917–1923, Vols. 1–3* (1950–1953); Treaty of Dorpat Between Finland and Russia, *LNTS* Vol. 3 (1921), Issue 1, No. 91.

34 Société des Nations, "Conférence Relative à la Non-Fortification et à la Neutralisation des Iles d'Aland," 10.

35 The original (and rejected by the Ålanders) Autonomy Act and the law for its implementation were adopted by the Finnish Parliament as Acts 124 and 125/1920.

36 Société des Nations, "Conférence Relative à La Non-Fortification et à la Neutralisation des Iles d'Aland," includes the multiple drafts by Finland and Sweden as an appendix.

Most of the discussion concerned the role of the Council of the League of Nations in cases of an alleged violation of the treaty. Anzilotti and Kaeckenbeeck, from the League of Nations Secretariat, wrote a note on aspects of collective security. They argued that states parties should clarify whether states parties could act alone in case of an alleged violation of the treaty, even if the council had not pronounced on this issue.[37] The conference participants also discussed the issue of access of foreign military ships to the demilitarised zone. Jean Gout, French representative, underlined that a state can refuse access of military ships to its territorial waters, including in times of peace. Gout argued that even Great Britain, a power in favour of the freedom of navigation, required prior notification for military ships. The draft was consequently in line with international law and only added to it limitations on the number of foreign military ships that could be in the Åland demilitarised zone simultaneously. The German representative, von Baligand, underlined that the proposed provision meant that "the right of Finland to receive visits by foreign military vessels is limited to one single vessel for each state". The minutes of the conference were approved, and the final text of the convention was adopted at the sixth session of the conference, which took place on 20 October 1921. The reason for this detailed account of the deliberations is to show that the 1921 Convention was the result of an important effort involving many states, experts from both the diplomatic as well as military professions and from the League of Nations Secretariat. It was not a hasty or impulsive product, but rather the outcome of a conscious and engaged multilateral deliberation and negotiation which combined idealist and realist elements. This fact and the agreement's detailed content have made it the main legal point of reference concerning the demilitarisation and neutralisation of Åland, without this putting in question other valid legal obligations, prior or posterior to that date.

The 1921 Åland Convention consists of ten articles providing mainly the following:[38]

- Finland, whose sovereignty over the islands had just been confirmed by the League of Nations, undertook not to fortify the islands;
- The geographic area of the demilitarised and neutralised zone was defined through a set of co-ordinates;

37 Ibid., 53.

38 For more detailed analysis of the 1921 provisions see Hannikainen (1994); H. Rotkirch, "The Demilitarization and Neutralization of the Åland Islands: A Regime 'in European Interests' Withstanding Changing Circumstances," *Journal of Peace Research* 23, No. 4 (1986) 357–76; P. Kleemola-Juntunen, *Passage Rights in International Law: A Case Study of the Territorial Waters of the Åland Islands* (2014); S. Spiliopoulou Åkermark, "Åland's Demilitarisation and Neutralisation: Continuity and Change," in *The Åland Example and Its Components: Relevance for International Conflict Resolution* (2011) 50–71; Björkholm and Rosas (1990); Söderhjelm (1928).

- The demilitarisation is regulated in detail and certain (limited) exceptions are admitted;
- The right to innocent passage was confirmed;
- The rules of neutralisation at times of aggression and war are set and the exceptions are defined;
- Guarantees for the convention's monitoring and maintenance were introduced.

Article 3 concerns and prohibits all kinds of military installations or installations that can be used for war purposes. Article 4 provides restrictions on the presence of military forces as well as the transport of military equipment and acknowledges the strict rights of Finland when responding to urgent needs of public security. Article 5 reconfirms the right of innocent passage under international law while Article 6 outlines the special conditions applying to situations of war, recognising anew the prime right and obligation of Finland to defend not only its territory but also the demilitarised and neutralised status of Åland. Article 7 sets the procedure to be followed by the League of Nations and Contracting Parties when there is a situation of aggression against the Åland Islands or against Finland through the islands. It created thus an early system of collective self-defence.[39] Article 8 included strong terminology according to which the rules set in the 1921 Convention shall remain valid whatever changes of circumstances may occur in the Baltic Sea.[40]

One of the most interesting—and often overlooked—provisions in the 1921 Convention is the one included in its Article 9. Here we find the expression of a commitment to bring the convention to the attention of all members of the League of Nations, whether parties to the convention or not, since the convention formed part of the valid rules directing governments ("*des règles de conduite effectives des gouvernements*") and it furthermore foresaw that contracting parties could invite other states to accede to the treaty, even though this did not occur.[41] Today, the UN Secretariat holds the role of the depositary for the 1921 Åland Convention. In 1992, Estonia registered with the UN Secretariat a declaration of continuity with regard to the position of Estonia as party to the convention. Estonia

39 S. Spiliopoulou Åkermark, "The Puzzle of Collective Self-Defence: Dangerous Fragmentation or a Window of Opportunity? An Analysis With Finland and the Åland Islands as a Case Study," *Journal of Conflict and Security Law* 22, No. 2 (2017) 249–74.

40 Hannikainen (1994).

41 Translation by the authors. See also Linderfalk, "International Legal Hierarchy Revisited: The Status of Obligations *Erga Omnes*," (2011) 1–23.

confirmed thus both its own commitment as well as the continued validity of the convention itself. This step was taken when the following factors led to the reshaping of politics and conditions of regional and of world politics, fuelling also domestic debates about the validity and meaningfulness of the demilitarisation regime: the dissolution of the USSR, the withdrawal of East Germany and Poland from the Warsaw Pact (prompting the dissolution of that pact), the unification of Germany and the deepening of European integration.[42]

2.3 The Åland Islands and Finland After the Second World War

After the signature of the Peace Treaty with Finland in 1947, the Åland Islands, according to Article 5 of the Treaty, was to "remain" demilitarised in accordance with the situation "as at present existing". The 1947 Peace Treaty for Finland included several limitations on the freedom of Finland to pursue an independent defence and security policy, with reference to Germany and its allies. On 6 April 1948, Finland and the Soviet Union signed the Treaty on Friendship, Cooperation and Mutual Assistance.[43] While confirming the principles of the newly adopted Charter of the United Nations, including non-interference in domestic affairs, the treaty obliged Finland to consult with the Soviet Union in situations of aggression and imposed a duty of co-operation and assistance between the countries. While retaining formal political independence, Finland had to pay war reparations and be in a sort of a loose alliance with the Soviet Union, something which prompted the usage of the noun 'Finlandisation' to describe a situation of nominal independence with considerable limitations for instance in the field of regional (Nordic and European) co-operation.[44]

Research on the fate of demilitarisation during the Cold War period has been limited partly due to the inaccessibility of archives until recently. This is slowly changing however, and more and more interest is being generated

42 UNTS, Vol. 1681, Annex C (255) 501–2. Y. Poullie, "Åland's Demilitarisation and Neutralisation at the End of the Cold War: Parliamentary Discussions in Åland and Finland 1988–1995," *International Journal on Minority and Group Rights* 23, No. 2 (2016) 179–210.

43 FTS 17/1948. See New Year speech by President Paasikivi in 1950.

44 K. Törnudd, *Soviet Attitudes Towards Non-Military Regional Co-Operation, Vol. 28: No. 1* (1961); R. Nyberg, "*Ni har vidrört VSB-avtalet*," in *Säkerhetspolitik och historia*, eds. K. Wahlbäck et al. (2007) 285–99.

on the multiple and varying experiences of what, in sum, were known as 'the Cold War'. For instance, the authors have been able to establish that in 1951, the British Foreign Office wrote a research note and an assessment of the legal position of the Åland regime as a response to a question received from the British embassy in Denmark. Miss J.A.C. Gutteridge, lawyer at the Foreign and Commonwealth Office (FCO), notes on this occasion that the 1921 Convention "is still in force" and that abrogation or modification of its provisions "would be possible only if all the parties thereto agreed that it should be terminated or that certain of its provisions should be modified".[45] Furthermore, the same note makes reference to Article 5 of the Treaty of Peace with Finland and to the view of the Second Legal Advisor at the time, Sir Gerald Fitzmaurice, according to which "both the provisions of Article 5 of the Finnish Treaty and our position as parties to the 1921 Convention would enable us to intervene if any question of the de-militarization of the Aaland Islands arose". Interestingly, the work and conclusion by Miss Gutteridge came to be reconfirmed much later by the FCO Legal Counsellor F.D. Berman. In 1981, Berman had to answer a request concerning confidentiality of the documents mentioned above. He approved disclosure and concluded:

> You might also like to bear in mind that it was *not* the 1921 Convention which neutralized the Aaland Islands, but a much earlier Agreement of 1856 (which you will see mentioned in the Preamble of the 1921 Convention). The new arrangement was necessitated (if that is the proper word) by changes of sovereignty in the Baltic, and it followed the consideration of the question by the Council of the League of Nations in 1920, a dispute having arisen between Sweden and Finland. There was a report from a Committee of Jurists appointed by the League Council, which is frequently cited in the international law textbooks as demonstrating that a territory may be subjected by treaty to a type of objective status which transcends the terms of the treaty itself as a contractual arrangement between its Parties alone and may accordingly be invoked by States not themselves Parties to the treaty. There is therefore a sense in which the neutral status of the Aaland Islands may be said to exist independently of the fate of the various treaties concluded to establish or confirm that status.[46]

45 Minutes dated 20 November 1951, signed by Joyce Gutteridge. National Archives, FCO 033/5081. Gutteridge is also the author of *The United Nations in a Changing World* (1969).

46 Emphasis as in the original. Confidential Note dated 23 July 1981, signed F.D. Berman, National Archives, FCO 033/5081.

Issues of international co-operation, defence and security policies were very much at the heart of Finland's precarious position during the Cold War.[47] The Soviet Union was even prepared to depart from the 1940 bilateral Åland treaty, if Finland would have accepted a closer military alliance with Moscow.[48] Over the years, Finland frequently reinterpreted the terms of the 1947 Peace Treaty, as was the case for instance in the summer of 1990, when the government proposed that a civilian helicopter for emergency services and rescue operations for Åland be ordered from the German company Messerschmitt-Bölkow-Blohm.[49] The Treaty of Friendship was a core element of societal and political debates around the year 1990. This period ended in a new bilateral agreement between Finland and the Russian Federation in 1992 replacing the 1948 Friendship Treaty.[50] Finnish ambassador René Nyberg has described this with the words "diplomacy is not law, but law can be diplomacy".[51] As mentioned earlier, the Russian Federation also signed that same year a Protocol with Finland concerning the succession by the Russian Federation with regard to bilateral agreements between Finland and the Soviet Union. The 1940 bilateral treaty on the Åland Islands was among those explicitly included.[52]

2.4 Times of Challenge

There have been many other occasions to challenge, reappraise and reconfigure Finnish international affairs with regard to the demilitarisation and neutralisation of the Åland Islands, as well as in the region and Europe as a whole.[53] There are likely to be many more in the future. It is more or less self-evident that times of war create pressures on a system of demilitarisation and neutralisation. During the First World War, Russia created fortifications which were notified—and accepted—as temporary by Great Britain and France, and also by Sweden, according to information available, after

47 Nyberg, "*Ni har vidrört VSB-avtalet*" (2007).

48 Nyberg, "*Ni har vidrört VSB-avtalet*" (2007), with reference to P. Visuri, *Puolustusvoimat kylmässä sodassa: Suomen puolustuspolitiikka vuosina 1945–1961* (1994).

49 Nyberg, "*Ni har vidrört VSB-avtalet*" (2007).

50 FTS 63/1992, *Överenskommelse med Ryska Federationen om grunderna för relationen mellan länderna*. The new Russian-Finnish agreement had been negotiated in January 1992 and was thereafter approved by the Finnish Parliament. It entered into force through Regulation 648/1992.

51 Nyberg, "*Ni har vidrört VSB-avtalet*" (2007).

52 *Ålands kulturstiftelse, Internationella avtal och dokument rörande Åland 1856–1992* (1993).

53 Issues of military exercises and of the handling of potential violations of the demilitarisation zone have been dealt with previously in Spiliopoulou Åkermark, "Åland's Demilitarisation and Neutralisation: Continuity and Change" (2011).

the increased presence of the German navy in the Baltic Sea.[54] After the war and many complicated turns and negotiations involving various states parties in changing constellations, the demilitarisation regime was reconfirmed and broadened mainly through two avenues. After the Finnish declaration of independence, a demolition commission was set up and negotiated the Agreement between Finland, Sweden and Germany on the demolition on fortifications on the Åland Islands.[55] The work of the League of Nations on the status and the demilitarisation of the Åland Islands completed this task with the 1921 Convention which widened both the scope of states parties as well as the material scope of the regime as discussed above. The Second World War resulted, as mentioned in the previous chapter, in the 1940 bilateral treaty between Finland and the Soviet Union, the 1944 Armistice Agreement and the 1947 Peace Treaty all of which confirmed the special status of the Åland Islands.

However, challenges have also arisen outside the immediate conduct of war and armed conflict in the strict sense. Not only did the Cold War differ from armed conflict. Two other situations illustrate this point and involve a multiplicity of actors at two different moments in history. The first is the demonstration (named 'Farmers' march', *bondetåget*) that took place in Mariehamn in the autumn of 1938 before the outbreak of hostilities in the Second World War. The second is the process for ratification of the Treaty on Open Skies by Finland in 2002. These two instances transgress the level of international law and politics if understood simply as actions and decisions at the interstate level. They show that security matters have had a profound and even identity shaping influence also on local understanding in the Åland Islands. Furthermore, from the autonomy theory perspective, the authors note that the Ålanders have been informed and engaged from an early stage in spite of the absence of formal legal competence in matters of security and defence.

The 'Stockholm Plan' and the 'Farmers' March'

Following the annexation of Austria by Nazi Germany in the spring of 1938, military and diplomatic circles in Finland and Sweden started communicating concerning a partial modification of the 1921 Convention and a limited remilitarisation of certain locations of the Åland archipelago.[56] The core points of this so-called Stockholm Plan were the exclusion of the southern

54 Söderhjelm (1928) 112–13 & 123–8.

55 FTS 152/1919.

56 League of Nations, "Official Journal," May–June 1939 and *Report on the Work of the League 1938/39*, 12.08.1939, Series of League of Nations Publications General 1939.2,

part of the archipelago (Lågskär, Björkör, Kökar) from the demilitarisation obligations, while the neutralisation would continue to apply in the entire zone. Sweden would contribute to the defence of the islands in a situation of war also with troops if needed.[57] The plan also proposed the introduction of conscription on the Åland Islands with Swedish as the language of command. After lengthy domestic discussions, the two governments proposed to the parties to the 1921 Convention in January 1939 a relaxation "authorised in virtue of Article 7, paragraph 1, of the 1921 Convention". At the same time, they also communicated these provisions to the government of the USSR. In May 1939, the issue was discussed by the Council of the League of Nations; the representative of Belgium and rapporteur for the issue, Maurice Bourquin, noted in his report that the signatory powers to the 1921 Convention had come to a "general agreement" on the application of the measures while the USSR delegate had proposed a postponement of the issue since his government did not consider itself "to be as yet in possession of all the information necessary to determine its attitude to a question which is of special concern to the Union of Soviet Socialist Republics as a Baltic Power in view of the geographical position of the Åland Islands".[58] The plan created lengthy and heated political debates in the Finnish and Swedish Parliaments as well as in the press in both countries. Eventually and also as a result of Soviet objections, the proposed plan was not implemented, so there was no modification of the existing conventional obligations regarding the demilitarisation and neutralisation of the Åland Islands. In this sense, the process entailed a reconfirmation of their demilitarisation and neutralisation regime and the commitment of all states parties to keep to its norms unless there was full agreement to deviate from them.

Meanwhile, the deteriorating situation in Europe and the autonomy legislation for Åland, which was seen as requiring revision, were also hot topics on the islands.[59] There followed, perhaps for the first time in the history of the autonomous region, a wider public debate about the demilitarisation and neutralisation regime of the islands.[60] The 1938 demonstration, referred to

---

21–3; K. Wahlbäck, *Finlandsfrågan i svensk politik 1937–1940* (1964); Stjernfelt (1991) 65–97; Hannikainen (1994) 621–3; Gustavsson (2012) 236–8.

57 Wahlbäck, *Finlandsfrågan i svensk politik 1937–1940* (1964) 385 concerning Östen Undén's positions to the issue of troops and the Åland Convention.

58 *Report on the Work of the League 1938/39*, 22–3.

59 S. Spiliopoulou Åkermark et al., "Untying a Sailor Knot" (Forthcoming article).

60 *Ålandstidningen* had however on its first page an article titled *"Ålänningarna och värnplikten"* (The Ålanders and Military Service) on 14 January 1920 referring to Ålandic argumentations on the basis of the 1856 Convention dating back to 1878. Perhaps this article could be attributed to Julius Sundblom, at the time chief editor of the newspaper and at the same time key political figure on the islands.

as the 'Farmers' March'—which alludes to the 1809 resistance on Åland towards the then-new Russian rule in Finland, including on Åland—brought together the issues of conscription provisions in the Autonomy Act of Åland (presently Act on the Autonomy of Åland [1991/1144]) and the international legal regime of the demilitarisation and neutralisation of the islands. According to one estimate, some 4 000 persons demonstrated in Mariehamn on 31 October 1938 against the two-sided threat of the proposal for military conscription in the (Finnish) army and of a modification or exception to the system of demilitarisation and neutralisation of the islands.[61] As was the case in the period preceding the 1921 decision of the League of Nations, and also in 1938, the Ålanders wanted to make their voice heard on matters on which they did not have formal legislative competence, but which were of considerable relevance to them. The resistance to conscription as well as to the modification of the demilitarisation and neutralisation regime, which could have meant a Swedish military presence on Åland, do not seem to signify a radical pacifistic line of action locally. Thousands of Ålanders joined a newly created local home guard, while many Ålanders also joined the Finnish army in the Winter War and the Continuation War, even though war activities on Åland were limited.[62] At the international law level, the right to self-defence thus appears also to be seen on Åland as running parallel to the valid, accepted and endorsed rules and agreements concerning the demilitarisation and neutralisation of the islands.

The Molotov–Ribbentrop Pact (officially the Treaty on Non-Aggression) between Germany and the USSR was concluded on 23 August 1939 and was followed by the invasion of Poland by Germany and the annexation of the Baltic States by the USSR.[63] Throughout the interwar period, Finland co-operated extensively with Germany on a wide range of issues, including military matters, sports and the planning of the Olympic Games scheduled for the summer of 1940 in Helsinki, and the establishment of a secret police in Finland.[64] The collapse of negotiations between Finland and the Soviet

61 *Ålandstidningen*, 1 November 1938. The march was also widely publicised in newspapers on mainland Finland as well as in Sweden. For a timeline of events see Gustavsson (2012) 236–8.

62 E. Tudeer, "Ålands hemvärn 1939–1940," *Särtryck ur krigshistorisk tidskrift*, No. 9 (1990). R. Nyberg, "Finlands säkerhetspolitik och Ålands status," *Hufvudstadsbladet*, 26 September 1990. Interestingly both of these publications appeared in 1990.

63 The secret protocols attached to the Molotov–Ribbentrop Pact were published much later, e.g. in Australian International Law News (1990) 77.

64 J. Lavery, "Finnish-German Submarine Cooperation 1923–35," *Scandinavian Studies* 71, No. 4 (1999) 393; E. Hübner, "Some Notes on the Preparations for the Olympic Games of 1936 and 1940: An Unknown Chapter in German–Finnish Cooperation," *The International Journal of the History of Sport* 30, No. 9 (2013) 950–62.

Union in November 1939 resulted in the Winter War between the two countries and in the exclusion of the USSR from the League of Nations at the international level by a decision of the Council on 14 December 1939 by virtue of Article 16(4) Covenant of the LoN. As discussed earlier, this stage of the war resulted in the 1940 bilateral treaty between Finland and the Soviet Union which also included provisions on the demilitarisation of the Åland Islands, but this time bilaterally.

The fact that approximately 20% of Åland's population demonstrated in Mariehamn and were addressed, among others, by Julius Sundblom, one of a number of influential politicians and at that time speaker of the Åland Parliament, who spoke strongly in favour of demilitarisation, was one of the first public discourses on matters pertaining to the status of Åland as a demilitarised and neutralised region.[65]

The Finnish Ratification of the Treaty on Open Skies

The end of the Cold War, the reunification of Germany, the disintegration of the USSR and the prospect of a broadened and deepened European integration resulted in profound changes for the political environment in and around Finland and the Åland Islands. While European integration is discussed in more detail in Chapter 4, it suffices here to note that in the early 1990s, there was both a sense of relief and optimism as well as a sense of considerable concern and insecurity.[66] In the case of the Åland Islands, there was for instance at the time another succession of persons, coming mainly from military circles in Finland, arguing about the dangers, risks and even meaninglessness of the demilitarisation and neutralisation of the islands; at the same time, international lawyers were writing about the effects of such transitions, concluding that overall international legal obligations concerning Åland had remained valid.[67] The increased presence of Finnish military vessels and aircraft in the Åland Islands archipelago also prompted reactions both from the Ålandic public as well as politicians.[68]

65 *Ålandstidningen*, 1 November 1938 and 3 November 1938 as well as an interview with the author Johannes Salminen in *Ålandstidningen* 20 August 2010 on his recollections as one of the youngest participants in the 1938 demonstration.

66 M. Jakobson, *Finland in the New Europe* (1998).

67 Summarised in Hannikainen, "The Continued Validity of the Demilitarised and Neutralised Status of the Åland Islands" (1994) 615–16; Poullie, "Åland's Demilitarisation and Neutralisation at the End of the Cold War—Parliamentary Discussions in Åland and Finland 1988–1995" (2016).

68 M. Koivisto, *Witness to History: The Memoirs of Mauno Koivisto, President of Finland 1982–1994* (1997).

In this mixed, fluid and volatile atmosphere, states of the Organization for Security and Co-operation in Europe (OSCE) were able to negotiate the Treaty on Open Skies. This treaty created a system of confidence-building measures through regular unarmed overflights for the purpose of observation from Vancouver (Canada) to Vladivostok (Russia), in other words in the entire area of the OSCE. The treaty was adopted in Helsinki in 1992 and entered in force on 1 January 2002, even though early efforts had already been made four decades earlier with special emphasis on air inspection in the Arctic.[69] The governments of Canada and Hungary hold the role of depositaries of the treaty following its Article XVII. Currently, thirty-four states have ratified the Treaty on Open Skies.[70] The agreement had originally been negotiated between the two military alliances, NATO and the Warsaw-pact countries. Finland and Sweden had been active in the negotiations in spite of not being members of either of the two alliances and they later participated as observers. Finland made considerable and largely successful efforts to influence also the substantive content of the treaty, for instance with regard to limiting 'spill-over information' in the vicinity of borders as well as limiting information gathered to and from the sites of inspection.[71] While this was perhaps not done with only, or even largely, Åland in mind, these efforts limit any possible negative effects for the Åland Islands demilitarisation and neutralisation regime. In November 2001, both countries (Finland and Sweden) announced their co-ordinated intention to accede to the Treaty on Open Skies but only after it had entered into force in accordance with its provisions.[72] According to Article XVII(2) this would first happen 60 days after ratification by all major military powers who had been allocated considerable flight quotas, which included, in addition, in Annex A of the treaty, "the Republic of Belarus and the Russian federation group of States Parties". The ratifications of Belarus and of the Russian Federation were only registered on 2 November 2001.

69 Text of the Treaty on Open Skies is available at the OSCE Document Library. See also P. Kaukoranta, "Negotiation of the Treaty on Open Skies Revisited: Finnish Features," in *Nordic Cosmopolitanism: Essays in International Law for Martti Koskenniemi*, eds. J. Petman and J. Klabbers (2003) 371–89; J.H. Spingarn, "Five Months in London," *Bulletin of the Atomic Scientists* 13, No. 7 (September 1957) 257–61. Spingarn was not only a lawyer but also a lieutenant commander in the US Naval Reserve.

70 See "Organization for Security and Co-operation in Europe: Treaty on Open Skies," 24 March 1992.

71 Kaukoranta, "Negotiation of the Treaty on Open Skies Revisited: Finnish Features" (2003).

72 H. Rotkirch, "A Peace Institute on the War-Path: The Application of the Treaty on Open Skies to the Neutralized and Demilitarized Åland Islands and the Powers of the Åland Autonomy," in *Nordic Cosmopolitanism: Essays in International Law for Martti Koskenniemi*, eds. J. Petman and J. Klabbers (2003) 61–88.

The Finnish Government considered that the purpose of the Treaty on Open Skies, which was transparency in military affairs and the use of confidence-building measures to uphold peace, coincided with the core purpose of the demilitarisation and neutralisation of the Åland Islands. As a result, it asserted that there was no contradiction between the treaty and the demilitarisation and neutralisation of Åland. An ambassador from the United States to the Open Skies conference even suggested that overflights could be employed to monitor respect of demilitarised zones.[73]

The issue of the ratification of the Treaty on Open Skies created considerable discussion on the Åland Islands, in the Åland Parliament as well as in the media. Information about the Finnish Government's intention to ratify the treaty reached the Åland Government and Åland Parliament on 30 November 2001. The Åland Parliament Committee competent on matters related to the autonomy and international matters (*Självstyrelsepolitiska nämnden*, or the Autonomy Committee) dealt extensively with the issue and heard a number of experts in the winter of 2002, including international lawyers.[74] In its opinion dated 10 April 2002, the Autonomy Committee explained its restrictive view on military presence in the Åland Islands as well as the importance of sound communication on international affairs between the Ålandic authorities and the government and parliament in Helsinki. The committee reflected along three main possible and not mutually exclusive avenues, leaving the final outcome open. One option was the usage of a territorial reservation with regard to the application of the Treaty on Open Skies on Åland; the other was information to the states parties concerning demilitarisation and neutralisation in conjunction with the Finnish ratification process and an emphasis on the strict interpretation of these international legal obligations; the third line of reasoning was that Finland should notify states parties that overflights were superfluous over the Åland Islands given there were no military installations to monitor.

In the process of parliamentary approval of the ratification of the treaty, the Finnish Parliament Foreign Affairs Committee concluded that there was no direct contradiction between this treaty and the Åland Convention, but nevertheless recommended that the Governor of Åland (*landshövding*) should be kept informed of the situation of overflights. The assumption was that the occurrence of overflights over the Åland Islands was unlikely.[75]

73 Ibid., 77.

74 Ålands lagting, "*Självstyrelsepolitiska nämnden*," Förslag nr 1/2001–2002.

75 Finnish Government Bill RP 71/2002 rd and UtUB 23/2002 rd, 15 October 2002. The avoidance of any such overflights was orally confirmed much later by Anders Gardberg on 7 September 2010 upon the occasion of a seminar on demilitarisation and neutralisation that took place in Mariehamn. Finnish Ministry for Foreign Affairs, "*Ålands demilitarisering*

Sweden's ratification was registered on 28 June 2002, while the Finnish ratification could finally be registered on 12 December 2002. The discussion around the Open Skies confirmed therefore that military presence, beyond the exceptions specified in the conventions or accepted by all states parties, is precluded irrespective of the proclaimed goal of such activity.

2.5 Remarks on the Legal Validity of the Norms

The above narrative shows the persistence of the demilitarisation and neutralisation regime of the Åland Islands despite challenges and deviations presented by the multiple layers of both world wars, considerable shifts in political and geopolitical circumstances such as the establishment of the League of Nations and later on of the United Nations, the Cold War period and its various stages, the end of the Cold War and the deepening of European integration to which Finland and Sweden have been highly involved.

The rules on demilitarisation and neutralisation entail restrictions in the limited geographical area called 'the demilitarised zone'. As a more recent reflection of this insight of continuity and restraint in the Åland case, one can mention the following example: in 2005–2013, the Finnish authorities co-operated with the Ålandic government in updating the cartographic co-ordinates of Åland demilitarisation and neutralisation.[76] This Åland solution is therefore not an encompassing system covering the Baltic Sea or the Nordic region as a whole. What happens in adjacent regions has consequences on the Åland Islands. Deviations are, however, restricted and regulated particularly by the provisions of the 1921 Convention with regard to the number and tonnage of foreign ships that can be authorised by Finland, the exercise of innocent passage, the presence of Finnish armed forces and the measures that can be permitted in a situation of imminent threat or war against Finland. Rather than competing in military presence and territorial control, states parties accepted on various occasions over time not to intervene in a limited, but not simply symbolic, territorial zone and to create a platform for communication around an issue which had

---

och neutralisering mot bakgrund av de nya europeiska utmaningarna [Åland Demilitarisation and Neutralisation Against the New European Challenges]," 2010.

76 Finnish Ministry of Justice, "*En utredning om gränserna för Ålands demilitarisering*," 2006. See also the summary of outcomes by the Ministry of Foreign Affairs which included the notification of states concerned. The notification was extended not only to parties in a treaty relation with regard to the demilitarisation, but in fact also to all countries around the Baltic Sea. The Circular Note to states concerned (HEL 7874–8 of 26 February 2013) has been published in the Handbook of the Åland Government on demilitarisation issues: Government of Åland, "*Policy för Ålands demilitarisering och neutralisering, Handbok för landskapets myndigheter*" (2015) 57.

been earlier disputed and controversial. This, one could conclude, is an early confidence-building measure and a precursor of a localised collective security regime.

Demilitarisation and neutralisation can be understood as a limitation to territorial sovereignty, in this case for Finland, since access to this territory is limited as far as military presence is concerned and taking into account that the limitations are harsher on other states than on Finland, in particular in times of war. Nonetheless, at the same time, the purpose of these rules is to confirm the idea of territorial sovereignty and control of territory. In fact, the Åland Islands regime is premised upon clear—but regulated—territorial sovereignty and thus the ability and legal right, as well as the obligation for Finland to repel attacks and imminent threats towards or through the zone in order to safeguard the demilitarised and neutralised status as well as its own territorial integrity. However, this very solution is an exception and a provocation to our thinking about the ways in which such territorial sovereignty can be exercised. This regime requires transparency and communication on alleged controversies, something which became even clearer in the 1940 bilateral treaty between Finland and the Soviet Union. Demilitarisation and neutralisation form perhaps a small and partial step towards disarmament and recognition of the fact that the arms race that took place in many countries prior to both 1914 and 1939 was a strong contributing factor to the outbreak of the devastating wars.[77]

In the following chapter, we shall examine closer the intersection between the law of the sea and the demilitarisation and neutralisation of the Åland islands.

77 Spiliopoulou Åkermark, "The Puzzle of Collective Self-Defence" (2017).

3 The Law of the Sea and the Demilitarisation of Åland

In May 1976, during Queen Elisabeth II's visit to Finland, two British warships escorted Her Majesty's ship *Britannia*. The warships had to remain outside the demilitarised and neutralised zone while the Queen visited Mariehamn, capital of the Åland Islands. A new issue arose in 1988, when foreign sailing ships taking part in the Tall Ships Race visited Mariehamn. The Finnish Government held then that the interdiction to permit more than one foreign warship to visit the zone did not apply to these unarmed sailing ships, and besides, the ships entered the zone solely with the aim of participating in a peaceful sailing contest. More recently, we know that Finland granted permission for four government ships and one warship to visit the demilitarised area in the period 2006–2013.[1] However, in August 2017, a negative decision was taken by Finland to allow a Russian school sailing ship, *Kruzenstern*, with 164 cadets on board, to make a visit to the port of Mariehamn. The Finnish Defence Command did not justify the decision.[2] During the autumn of 2003, the ferry *Silja Europa* departed from Stockholm and called at the port of Långnäs in the Åland Islands, following its regular route between Stockholm and Helsinki. On this particular occasion, however, in addition to regular passengers, the ferry also transported Swedish troops along with their vehicles and weapons, on their way to a Nordic Peace military exercise held in Finland. This incident, which included several other complicating circumstances, prompted several reactions at domestic and regional level.[3]

1 Information received from the Finnish Ministry of Defence on visits between 2006 and 2013, 24 August 2017. The number of possible rejections is, however, not visible in the information sent by the authorities.

2 M. Gestrin-Hagner, "*Huvudstaben förbjöd ryskt skolfartyg att besöka Åland*," *Hufvudstadsbladet*, 28 August 2017.

3 Finnish Ministry of Defence, "Press Release No. 65–2003," Nordic Peace 2003—*Harjoitus ja Ahvenanmaa*, 12 September 2003. For a comment see Spiliopoulou Åkermark, "Åland's

These examples illustrate many intricacies related to the rights of passage under the current Åland regime and an evolving law of the sea. The principle international convention concerning the demilitarisation and neutralisation is nearly a century old; it regulates how ships are to navigate within the waters surrounding the Åland Islands and includes minimal provisions regarding airspace and other forms of military activities. Since the comprehensive evolution of the law of the sea since the entry into force of the 1921 Åland Convention, it also influences the conduct of navigation within these waters. The purpose of this chapter is to discuss the application of the law of the sea in the territorial waters of the Åland Islands and the Åland Strait in light of the 1921 Convention on the Non-Fortification and Neutralisation of the Åland Islands. This chapter, technical as it may seem in nature, carries evidence of the importance that has been attached over time to thorough legal reasoning in the interpretation of the multiple relevant rules.

Finland did not make any reservations to the 1958 Convention on the Territorial Sea and the Contiguous Zone (TSC) in terms of demilitarisation or neutralisation. The 1982 UN Convention on the Law of the Sea (LOSC) does not allow for reservations; only declarations are possible.[4] Finland and Sweden submitted such declarations when signing and ratifying the LOSC only with reference to passage through the Åland Strait.[5]

The preamble to the LOSC states that the convention has been elaborated in order to settle all issues relating to the law of the sea and to establish

> a legal order for the seas and oceans which will facilitate international communication, and will promote the peaceful uses of the seas and oceans, the equitable and efficient utilization of their resources, the conservation of their living resources, and the study, protection and preservation of the marine environment.

The preamble to the 1921 Åland conventions says that the convention has been made in order to guarantee peace and stability, in the sense that the

---

Demilitarisation and Neutralisation: Continuity and Change" (2011). See also P. Kleemola-Juntunen, *Passage Rights in International Law: A Case Study of the Territorial Waters of the Åland Islands* (2014) 1–2.

4 1833 UNTS 397; *Lag om godkännande av vissa bestämmelser i Förenta Nationernas havsrättskonvention och i avtalet om genomförande av del XI i den*, Statutes of Finland 524/1996, *Förordning om ikraftträdande av Förenta Nationernas havsrättskonvention och av avtalet om genomförande av del XI i den samt av lagen om godkännande av vissa bestämmelser i konventionen och i avtalet*, Statutes of Finland 525/1996; the text of the Treaty is published in FTS 50/1996.

5 See United Nations Treaty Collection, "United Nations Convention on the Law of the Sea, Montego Bay, 10 December 1982—Declarations".

Åland Islands shall never become a threat from a military point of view. Hence, it seems that, according to the preambles, the conventions are not contradictory.

As regards sea areas, the 1921 Åland Convention seeks to achieve its aims through Article 4 that restricts access of foreign warships and Finnish warships to the demilitarised zone and Article 5 that grants the right of innocent passage for foreign warships through the zone. The right of innocent passage for warships has been a controversial issue within the context of the law of the sea. States' opinions have been divided over issues concerning prior authorisation in respect of an unrestricted right of passage.[6] The content of provisions included in the 1921 Åland Convention regarding entering and anchoring are not particularly exceptional. The provision concerning the right of innocent passage, on the other hand, appears to be unusual. Neither the TSC nor the LOSC include a provision that specifically deals with the right of innocent passage of warships. Provisions regarding the right of innocent passage are situated under the sub-section covering all ships, and several states have interpreted that also warships enjoy the right of innocent passage subject to prior notification/authorisation. The provisions of the LOSC (in total 320 Articles, including the annexes, which form an integral part of the convention) include sections dealing with the territorial sea (Part II) and with navigation through straits lying partly or wholly within the territorial sea (Part III), and overlap to some extent with the provisions of the 1921 Åland Convention. All these conventions exist thus in parallel. Article 311(2) deals with the relation of the LOSC to other conventions. Paragraph 2 says that "[t]his Convention shall not alter the rights and obligations of States Parties which arise from other agreements compatible with this Convention and which do not affect the enjoyment by other States Parties of their rights or the performance of their obligations under this Convention". The provision refers to existing and future conventions. This suggests that the LOSC is given precedence over other conventions in the sense that it provides a yardstick against which the compatibility of other conventions is to be measured.[7]

6 For additional information on state practice, see Y. Tanaka, *The International Law of the Sea* (2015) 91–3; J.A. Roach and R.W. Smith, *Excessive Maritime Claims*, 3rd edn. (2012) 250–1; D.P. O'Connell, "Innocent Passage of Warships" (1977) 7 *Thesaurus Acroasium, The Law of the Sea (4th session: September 1976)* 408–51 at 413–16. For a list of states restricting innocent passage see W.K. Agyebeng, "Theory in Search of Practice: The Right of Innocent Passage in the Territorial Sea" (2006) 39 *Cornell International Law Journal* 371–99 at 396–8.

7 M.H. Nordquist et al., *United Nations Convention on the Law of the Sea 1982*. Volume V (1989) 243.

In the following sections, we briefly present the tools used in the law of the sea: legal rights, maritime zones, and navigational rights affecting the sea around the Åland Islands.

3.1 Maritime Jurisdictional Zones

Article 2 II of the 1921 Åland Convention stipulates:

> The territorial waters of the Aaland Islands are considered to extend for a distance of three marine miles from the low-water mark on the islands, islets and reefs not permanently submerged, delimited above; nevertheless, these waters shall at no point extend beyond the lines fixed in paragraph I of this Article.

The low-water mark refers to the low-water line, which was then an established rule of customary international law. In the second paragraph of Article 2, the convention defines the method to be used when demarcating the "territorial waters" of the Åland Islands. The first paragraph of Article 2 demarcates a demilitarised and neutralised zone by enclosing all land areas within a given range of co-ordinates. According to the co-ordinates provided, the line drawn from them defines *how far the demilitarised and neutralised sea area may extend.* Thus, the 1921 Åland Convention states that the demilitarised and neutralised zone shall not exceed three nautical miles beyond the co-ordinates provided.

The 1921 Åland Convention uses the term 'territorial waters'. The phrasing of Article 2 indicates that the term 'territorial waters' refers to the sea area, which is presently known as 'territorial sea'.[8] The 1920 Treaty of Dorpat defined the breadth of the Finnish territorial sea as four nautical miles (nm). The 1921 Åland Convention established the three nm territorial waters around the islands. Historically, the breadth of territorial sea pertaining to the Åland Islands, as defined by the 1921 Convention, has never been consistent with the breadth of Finland's territorial sea area. At the time of finalising the 1921 Åland Convention, Finland's territorial sea breadth was four nm. Finland suggested that four nm would have been the breadth of Åland's territorial waters too, which would have followed Danish, Norwegian and Swedish practices, as they each had territorial seas with such breadth. However, other parties negotiating the 1921 Convention did not endorse a four nm territorial sea. They referred to the narrowness of the Gulf of Bothnia,

8 J.O. Söderhjelm, *Démilitarisation et neutralisation des Iles d'Aland en 1856 et 1921* (1928) 247–8.

claiming that it was not convenient to expand the zone more than necessary. This was in fact not so surprising, as many Nordic countries had previously experienced similar opposition when establishing their territorial seas. Non-Nordic countries had not been sympathetic to claims for four-nm-wide territorial seas.[9] The three-mile limit gained support towards the end of the 18th century and retained its popular status throughout the 19th century. It was in the best interests of all major naval powers to ensure that territorial seas were not too wide. Jessup, for instance, considered the three-mile limit to have emerged as a general rule in international law.[10]

When Finland defined its sovereign maritime zones in internal legislation for the first time in 1956, it chose to use the method of straight baselines.[11] The establishment of straight baselines was also then part of customary international law and was codified in the TSC and later in the LOSC.[12] The breadth of Finland's territorial sea remained at four nm and base points were not allowed to exceed twice the breadth of the territorial sea. In the Åland Strait, the baselines had the effect of enclosing the area east of the Märket Reef as internal waters. As a consequence, the legal status of this sea area changed as the part of the Åland Strait belonging to the demilitarised and neutralised area became Finland's internal waters; this influenced the sea range area where the right of innocent passage applied.

Finland has twice defined its sovereign maritime zones and their borders in its legislation, in 1956 and then again in 1995.[13] This meant, as far as the Åland Islands area was concerned, that Åland Strait is not only the maritime border but also forms a country border between Finland and Sweden. Since the 1809 Treaty of Fredrikshamn, the borderline between Sweden and Finland (as a Grand Duchy of the Russian Empire) has been located in the Märket Reef. Therefore, Märket is in fact only partly demilitarised and neutralised, because the Swedish half is not covered by the conventions. Today the navigational route through the Åland Strait traverses Swedish

9 "Conférence Relative à la Non-Fortification et à la Neutralisation des Iles d'Aland," *Actes de la Conférence*, 30–1; P.C. Jessup, *The Law of Territorial Waters and Maritime Jurisdiction* (1927) 24, 31–41.

10 Jessup, *The Law of Territorial Waters and Maritime Jurisdiction* (1927) 7.

11 Act on the Delimitation of the Territorial Waters of Finland 18 August 1956; Statutes of Finland 463/1956, amended on 5 March 1965; Statutes of Finland 144/1965, 25 March 1966; Statutes of Finland 332/1966, 5 December 1969; Statutes of Finland 781/1969, 3 March 1995; Statutes of Finland 981/1995 entered into force on 30 July 1995 as enacted by the Decree of 17 July 1995; Statutes of Finland 982/1995.

12 *Fisheries* case, International Court of Justice Reports 1951 (United Kingdom v Norway).

13 Act on the Delimitation of the Territorial Waters of Finland 18 August 1956; Statutes of Finland 463/1956; Statutes of Finland 144/1965; Statutes of Finland 332/1966; Statutes of Finland 781/1969; Statutes of Finland 981/1995; Statutes of Finland 982/1995.

territorial sea and is located between the Understen islet on the Swedish side and the Märket Reef.[14] Although Finland's maritime boundaries have changed over time, the breadth of the demilitarised and neutralised sea area has remained intact.

Finland stated during the negotiations of the 1921 Convention that this sea area was meant to set the limits for the demilitarised and neutralised sea area; it did not affect Finland's right as a coastal state to establish maritime zones. As result of the new straight baselines established in 1995, most of the demilitarised and neutralised sea area is thereby part of Finland's internal waters today. At that time Finland also widened its territorial sea from four nm to twelve nm, the maximum breadth following the LOSC. This means that these sea areas are under Finland's sovereignty; sovereignty of a coastal state extends to the territorial sea and the airspace over the territorial sea. Sovereignty entails that the coastal state's laws are applicable in the territorial sea and that enforcement measures are reserved for the coastal state. Coastal states are obliged to exercise their sovereignty in conformity with the provisions of international law.[15] It is also noteworthy that the LOSC, which entered in force in 1994, does not set the minimum breadth of the territorial sea, but it may nowadays be claimed that the twelve-mile limit is firmly established in international law.

The extension by Finland of the territorial sea to twelve nm has some exceptions with regard to the areas of the Gulf of Finland, the Sea of Åland and the Gulf of Bothnia, where such extension was not possible. As mentioned above, Finland has applied the straight baseline method since the 1950s. According to previous legislation, base points were not allowed to exceed twice the breadth of the territorial sea. However, when the Act on the Delimitation of the Territorial Waters of Finland was changed in 1995, this kind of special limitation was not considered necessary.[16] Neither the TSC nor the LOSC limit the distance between base points. Limitations on the distance between base points are in fact not common in the practice of other states. Finnish legislation on the delimitation of territorial waters includes an obligation to update base points every 30 years due to continental uplift.[17] Finland, however, needed to consider agreements made with

14 C.N. Gregory, "The Neutralization of the Aaland Islands," *American Journal of International Law* 17 (1923): 63–76. Rotkirch (1986) 359.

15 *Yearbook of the International Law Commission*, Volume II (1956) 253, 265.

16 Act on the Delimitation of the Territorial Waters of Finland 18 August 1956; Statutes of Finland 463/1956, amended on 5 March 1965; Statutes of Finland 144/1965, 25 March 1966; Statutes of Finland 332/1966, 5 December 1969; Statutes of Finland 781/1969, 3 March 1995; Statutes of Finland 981/1995, entered into force on 30 July 1995 as enacted by the Decree of 17 July 1995; Statutes of Finland 982/1995.

17 Ibid. See also section 2.5 of this book.

other states before expanding the breadth of its territorial sea. For instance, the Åland Sea was one of these special maritime areas. While Finland was preparing to widen its territorial sea, it began negotiations with Sweden regarding the maritime boundary pertaining to the Åland Sea and the northern part of the Baltic Sea. The Agreement Concerning the Delimitation in the Åland Sea and the Northern Part of the Baltic Sea of the Finnish Continental Shelf and Fishing Zone and the Swedish Economic Zone follows of the Decree on the Application of the Act on the Delimitation of the Territorial Waters of Finland, which states that Finland's outer boundary within the Åland Sea area does not, at any time, cross into Sweden's Exclusive Economic Zone (EEZ).[18]

Bogskär, located in the southern part of the Åland Islands and consisting of two islets and their surrounding three-nm zone, has been a problematic area for Finland and Sweden when establishing maritime boundaries.[19] As early as the 1920 Treaty of Dorpat, the breadth of territorial waters around Bogskär differed from the general breadth of Finland's territorial waters. According to Article 3(3) of the treaty, Bogskär had consisted of three nautical miles of territorial waters because it was situated outside of Finnish territorial waters.[20] When Finland began using the straight baselines method in 1956, the breadth of territorial waters remained unchanged at four nm, while Bogskär kept its three-nm territorial waters. Nevertheless, after Finland expanded its territorial sea to twelve nm and the restrictions on the maximum distance between base points ceased, the Finnish territorial sea became contiguous to include Bogskär. Bogskär retains its unique character today as a distant, uninhabited place, outside of the straight baselines system. From this reasoning, it follows that Finland's territorial sea to the south of Bogskär does not extend beyond three nm.[21]

18 Section 5a (2), Statutes of Finland 981/1995; Finnish Government Bill HE 114/1994 vp on proposed changes of legislation pertaining to the Act on the Delimitation of the Territorial Waters of Finland and acceptance of treaty arrangements pertaining to delimitation of territorial waters, continental shelf and fishing zone of Finland (*Hallituksen esitys eduskunnalle laiksi Suomen aluevesien rajoista annetun lain muuttamisesta sekä Suomen aluevesien, mannermaajalustan ja kalastusvyöhykkeen rajoja koskevien sopimusjärjestelyjen hyväksymisestä*).

For the first time ever, the inner limits of internal waters were defined. This was due to a change in scope of the 1992 Convention on the Protection of the Marine Environment of the Baltic Sea Area (in force since January 2000).

19 The island of Bogskär is located approximately twenty-four nm south of the main island of Åland.

20 Treaty of Dorpat, Statutes of Finland 20–21/1921.

21 Finnish Government Bill 114/1994.

The straight baselines method also covers the Åland Islands area. A part of the Åland Islands' territorial waters is located in a zone that is also a part of Finnish internal waters. Straight baselines that were drawn up in 1995 enclosed parts of Finland's territorial sea and the high seas, and subsequently claimed these areas as internal waters.[22] The Act on the Delimitation of the Territorial Waters of Finland defines water areas that are part of internal waters and those that are part of Finland's territorial sea. This classification is also applied to the sea area surrounding the Åland Islands. The 1921 Åland Convention does not restrict Finland from defining maritime zones within the sea area of the Åland Islands, but it does specify the method to be used when measuring the demilitarised and neutralised three-nm zone. Hence, it does not prescribe maritime zones and their breadth around the islands.

Article 2 of LOSC says that "[t]he sovereignty of a coastal state extends, beyond its land territory and internal waters". In comparison to territorial seas, internal waters have a closer connection to land territory as a state has important interests to consider, such as territorial integrity, defence and commerce. In addition to land and sea, a state's territorial sovereignty extends to include the airspace above its internal waters and the sea-bed and subsoil areas below.[23] A baseline separates internal waters from territorial seas, acting as a dividing line between maritime zones and internal waters where foreign states have no general rights. Internal waters are not subject to regulation according to any of the Law of the Sea Conventions, because coastal states' territorial sovereignty over their internal waters is linked to the land domain and is therefore subject to the same legal regime.[24] However, despite the fact that coastal states enjoy a legal right to mark their territorial boundaries, international influence on sovereignty is inescapable because international law sets the rules for drawing baselines.[25] In international law, the difference between the legal status of internal waters and territorial seas is significant, as foreign ships do not enjoy a general right of innocent passage through internal waters. However, this rule is not without some exceptions based on treaty law. For example, after establishing straight baselines, sea areas may thereafter be considered internal waters,

22 In 1995, the straight baselines method enclosed as internal waters areas of the demilitarised and neutralised zone which had previously been considered as territorial sea. Discussion between P. Kleemola-Juntunen and Jukka Varonen, Finnish Transport Agency, 23 October 2013.

23 R. Lagoni, *Encyclopedia of Public International Law* 11 (1989) 153–5 at 153.

24 Ibid. See also R.R. Churchill and A.V. Lowe, *The Law of the Sea* (1999) 31, 60–1; N. Klein, *Dispute Settlement in the UN Convention on the Law of the Sea* (2009) 264; K. Hakapää, *Uusi kansainvälinen oikeus* (2010) 386.

25 Lagoni (1989) 155. This was pointed out by the ICJ in the 1951 *Fisheries* case.

although the right of innocent passage still remains through those waters.[26] Coastal states enjoy the same kind of sovereign rights over their internal waters as they do with their land territory. As mentioned, a state's territorial sovereignty extends to include the airspace above its internal waters and the sea-bed and subsoil areas below.[27] The choice of baseline method (normal or straight) is left to the discretion of a coastal state. Furthermore, a coastal state's jurisdiction over foreign ships differs within different maritime zones. In principle, a coastal state may enact laws and apply them to foreign ships within its internal waters.[28]

A coastal state's sovereign jurisdiction differs in regard to its internal waters, where no restrictions apply to the state's jurisdiction, and its territorial sea, as governed by the rules of international law.[29] It could be argued that the extension of the territorial sea to twelve nm had a positive effect, in that there is today a wider sea area around the demilitarised and neutralised zone where Finland may exercise necessary control over foreign ships to prevent or regulate military maritime activities and where overflight needs prior authorisation. The 1921 Åland Convention obliges Finland to stop or repel an attack against the demilitarised and neutralised zone; the extension of the territorial sea to twelve nautical miles enables Finland to respond to security threats earlier than was the case in the past and thereby guarantee better the respect of the demilitarisation and neutralisation regime and of the territorial integrity of the country.

3.2 Navigational Rights

With reference to navigational rights, the most important limitations to a coastal state's sovereignty relate to the right of *innocent passage* through the territorial sea, including a right of innocent passage of warships, and the rules concerning *transit passage*. As discussed, territorial sovereignty involves sovereign rights and exclusive jurisdiction. The 1958 and 1982 Law of the Sea Conventions do not include a general provision relating to the right of entry of foreign ships. Consequently, internal waters are left to customary international law regulation. A state has a right to decide what

26 TSC Art. 5 (2); UNCLOS Art. 8 (2).

27 Lagoni (1989) 153.

28 E.D. Brown, *The International Law of the Sea, Volume Introductory Manual* (1994) 38; I. Brownlie, *Principles of Public International Law* (2003) 299–300; H. Yang, *Jurisdiction of the Coastal State over Foreign Merchant Ships in Internal Waters and the Territorial Sea* (2006) 48.

29 Lagoni (1989) 153; K. Bangert "Internal Waters," *Max Planck Encyclopedia of Public International Law*.

kind of activities it allows inside its internal waters and whether prior permission to undertake such activities is required. In internal waters, a coastal state's sovereignty is, of course, not restricted by any general obligation to grant the right of innocent passage to foreign vessels.

The 1930 Hague Conference was more informative about warships' rights of innocent passage, as it stated that, "[a]s a general rule, a coastal State will not forbid the passage of foreign warships in its territorial sea and will not require a previous authorisation or notification".[30] This draft article reflects today's recognised practice in many states.[31] Although, as a widely accepted rule, coastal states may not require prior authorisation or notification for vessels entering their territorial seas, states do have a right to regulate conditions of passage and quite a few insist still on such authorisation or notification. This rule is particularly relevant when restricting the number of foreign vessels passing through coastal states' territorial seas at a given time or within a certain part of the sea.[32] The 1930 Hague Conference was held with the purpose of codifying rules of international law. It was successful in its work on territorial waters, but states could not reach an agreement on the breadth of territorial seas, so it failed to produce a convention.[33] Although the task of the conference was to codify existing rules of international law, participating delegates made it clear that they found it hard to draw any distinction between codification and making new international legislation. Article 12 on the passage of warships found in the Hague draft did not reflect the existing rules of customary international law.[34] Thus, it is worth wondering whether a new set of rules could have been established from the conference. The debates held at the 1958 and 1982 Law of the Sea Conventions do not seem to support the existence of this kind of rule of customary international law.[35]

30 Draft art. 12, R. Shabtai (ed.), *League of Nations, Conference for Codification of International Law* (1930) 1975, 1408.

31 W. Heintschel von Heinegg, "Warships," *Max Planck Encyclopedia of Public International Law, 2015.*

32 Shabtai (1930) 1975, 1418–19.

33 The only outcome of the conference was the Convention on Certain Questions Relating to the Conflict of Nationality Laws, 179 *LNTS* 89. However, during the *Corfu Channel* case, the United Kingdom referred to the majority of states' responses to the questionnaire circulated before the 1930 Hague Conference and 1921 Åland Convention as evidence of the right of innocent passage of warships through territorial seas; *Corfu Channel* case, I.C.J. Reports 1949, 43.

34 Documents on the Development and Codification of International Law, Part 3: The First Conference for the Codification of International Law (1947) 41 *American Journal of International Law*, Supplement, 83–5.

35 See debates about conflicts between developing countries and developed countries in M.H. Nordquist et al. (eds.), *United Nations Convention on the Law of the Sea 1982: A Commentary*. Volume II (1993) 195–9; and UNCLOS I, Official Records, Volume III, 129–31.

The number of draft proposals presented at the Third Conference on the Law of the Sea indicate that questions related to the passage of warships through territorial seas and straits used for international navigation have not been discussed since, and were still controversial after, more than forty years since the 1930 Hague Conference.[36] During the lengthy negotiations for the LOSC, coastal states pointed out that state sovereignty and security should be properly accounted for and should be safeguarded by enforcing prior authorisation or notification requirements upon warships, which should also observe coastal state rules and regulations.[37] Prior authorisation was, however, never intended to restrict states or vessels but was supposed to limit their intentions or policies.[38] The United Kingdom, one of the states opposing proposals for prior authorisation or notification requirements, disliked the notion that restrictions could apply to warships as well as hinder strategic, tactical and operational naval competencies.[39] Motions for and against rights of innocent passage for warships led the President of the Conference to stress that a 'package deal' was the aim of the whole conference.[40] The President came to a conclusion that stated that state delegations confirmed that no hindrance to coastal state rights and interests would be a result of the so-called package deal.[41]

Nevertheless, as Ngantcha writes, "it is clear that no true consensus was achieved".[42] Coastal states had included prior authorisation or notification requirements in their national legislation and stated that they found no incompatibility between their positions and provisions of the LOSC.[43] Some states even considered there to be an implicit recognition for prior notification or authorisation in the provisions of the convention. The delegate of Iran expressed an opinion of this kind when he clearly stated that,

36 See Malaysia, Morocco, Oman and Yemen: draft articles on navigation through the territorial sea, including straits used for international navigation, UN Doc. A/CONF.62/C.2/L.16, UNCLOS III, Official Records, Volume III, 194.

37 They found that this was necessary for the maintenance of peace, good order and security. See UNCLOS III, Official Records, Volume II, 142.

38 See UNCLOS III, Official Records, Volume II, 142.

39 A/CONF.62/C.2/L.3 United Kingdom: draft articles on the territorial sea and straits, UNCLOS III, Official Records, Volume III, 185.

40 For more on the 'package deal', see H. Caminos and M.R. Molitor, "Progressive Development of International Law and the Package Deal" (1985) 79 *American Journal of International Law*.

41 UNCLOS III, Official Records, Volume XVI, 132 (A/CONF.62/SR.176).

42 F. Ngantcha, *The Right of Innocent Passage and the Evolution of the International Law of the Sea: The Current Regime of "Free" Navigation in Coastal Waters of Third States* (1990) 145.

43 Out of the Baltic Sea states, Finland and Sweden in particular have said that a 'special regional rule' exists; see UNCLOS III, Official Records, Volume XVI, 52.

with reference to customary international law, the provisions of Article 21 read in conjunction with Article 19 on the meaning of innocent passage and Article 25 on the rights of protection for coastal states, which, combined, implicitly recognise the rights of coastal states to take measures to safeguard their security interests. Security interests may include the adoption of laws and regulations regarding the requirement of prior authorisation for warships intending to exercise the right of innocent passage through territorial seas.[44] Similar statements were made by other delegations during the signing ceremony, which reaffirmed the popular demand for prior authorisation or notification requirements for warships as they undertake innocent passage through foreign territorial waters. States agreed that this sort of requirement was consistent with existing international law and practice, which could be exercised on the basis of sovereignty, national integrity and state security.[45] This evidence illustrates the reluctance of coastal states to permit passage of warships without prior authorisation or notification. One reason for this disagreement concerns the fact that warships are often considered as representing a threat.[46] The development of armoury has made warships more effective and enhanced their roles in warfare. Coastal states have also become more alert to the threat of nuclear-powered ships and ships carrying nuclear weapons. Nevertheless, prior authorisation or notification requirements have not been included in treaty provisions, although they were requested by fifty states before adoption of the LOSC was made possible.[47] Maritime powers criticised claims made for prior authorisation or notification requirements, influencing also changes to Finnish and Swedish legislation. Sweden and Finland abolished their requirement for prior notification before ratifying the LOSC.[48]

The situation concerning the concept of innocence changed in 1982 following adoption of the LOSC. The convention includes specific definitions

44 UNCLOS III, Official Records, Volume XVI, 29–30; Volume XVII, 106 (A/CONF.62/SR.191).

45 UNCLOS III, Official Records, Volume XVII, 123–24; see also plenary meetings, Volume XVI, 20–52; also other plenary meetings.

46 As E. Root has already stated in 1910 during the *North Atlantic Coast Fisheries Arbitration*, in J.B. Scott (ed.), *Argument of the Honorable Elihu Root on behalf of the United States, Before the North Atlantic Coast Fisheries Arbitration Tribunal at The Hague, 1910* (1912).

47 F.D. Froman, "Uncharted Waters: Non-innocent Passage of Warships in the Territorial Sea" (1984) 21 *San Diego Law Review* 625–89 at 642; see UNCLOS III, Official Documents, Volume XIV, 15–81; Roach and Smith, *Excessive Maritime Claims* (2012) 240–51.

48 Hakapää, *Uusi kansainvälinen oikeus* (2010) 394; Government Proposal HE 12/1996 vp; Roach and Smith (2012) 250–1. The US protested against claims by many states that restrict the innocent passage of warships by requiring prior notification or authorisation. See Roach and Smith (2012) 243–59. See also the US Department of Defense (DoD), Freedom of Navigation, Report for Fiscal Year 2013.

concerning innocent passage. The right of innocent passage is regulated by Article 19, and paragraph 2 contains a list of activities that are considered to be prejudicial to the peace, good order or security of the coastal state. This convention also includes a specific rule for innocent passage of submarines, which is the same in content as its predecessor, the TSC. However, Article 23 goes further and sets obligations for foreign nuclear-powered vessels and vessels carrying nuclear or other inherently dangerous or noxious substances.[49] These more precise regulations concerning passage exist to interpret the concept of innocent passage, especially since this involves restrictions placed on the coastal state authority to enforce protective rules.[50]

It seems obvious that the aim of the LOSC was to produce a more objective definition that would leave coastal states less scope for interpretation, as well as less potential to abuse their rights when preventing non-innocent passage. There are particular references made to 'activities' in the text of the convention. Therefore, a vessel's presence or passage alone cannot be interpreted as prejudicial to coastal state interests if the vessel does not engage in some specific actions. Thus, at least in theory, the new formulation of the provision regulating innocent passage would widen the scope of the right of innocent passage. Any activity that has no direct bearing on passage will not automatically render passage non-innocent. Churchill and Lowe claim that activities seen as posing a threat of force affect third states as well as coastal states. In addition, because a coastal state's security is seen by Churchill and Lowe as being indirectly linked to a third state's welfare, they also believe that there is no need for links to other legal instruments, such as a mutual defence treaty, when aiming to render threats as incompatible with innocent passage. Thus, Article 19(2) may be interpreted in such a way as to allow coastal states to act on the impression that a third state's security is at stake.[51] Anand and O'Connell believe that to some extent, the list of activities presupposes a right of innocent passage for warships. In this sense, therefore, such activities would concern the mode of passage of warships as well.[52]

The United States and the former Soviet Union signed the bilateral Treaty on the Uniform Interpretation of Norms of International Law Governing

49 Art. 23: Foreign nuclear-powered ships and ships carrying nuclear or other inherently dangerous or noxious substances shall, when exercising the right of innocent passage through the territorial sea, carry documents and observe special precautionary measures established for such ships by international agreements.

50 UNCLOS, Arts. 21, 22; D. Pharand, "International Straits" (1977) 7 *Thesaurus Acroasium, The Law of the Sea (4th Session September 1976)* 64–100 at 77–8.

51 Churchill and Lowe, *The Law of the Sea* (1999) 85.

52 R.P. Anand, "Transit Passage and Overflight in International Straits" (1986) 26 *Indian Journal of International Law* 72–105 at 85; O'Connell, "Innocent Passage of Warships" (1977) 292.

Innocent Passage in 1989.[53] The Uniform Interpretation does not leave an understanding of innocence open to interpretation. Paragraph 3 of this treaty states that Article 19 of the 1982 Law of the Sea Convention sets out in paragraph 2 a list of activities that would render passage non-innocent. A ship passing through the territorial sea that does not engage in any of those activities is in innocent passage. The right of innocent passage is a clear exception to coastal state sovereignty. In other words, the right of innocent passage exists in direct relation to coastal state sovereignty over territorial waters. Other rights may be granted to foreign vessels according to a specific treaty, but the right of innocent passage exists independently of special arrangements. International law, however, gives coastal states the right to suspend innocent passage in certain circumstances,[54] which implies that the right of innocent passage is subordinate to coastal state sovereignty. A foreign vessel may be excluded from a territorial sea if its passage is non-innocent or moves beyond the scope of passage.[55] According to Article 25(1) LOSC, coastal states are allowed to take necessary steps to prevent non-innocent passage from taking place in their territorial seas.

Today, a majority of coastal states recognise the right of innocent passage of foreign warships, but such practice is not uniform. Currently, around forty states require prior notification or authorisation of the passage of warships through their territorial sea.[56] When Finland and Sweden signed the LOSC, they stated in their declarations that they require prior notification for the innocent passage of warships. This practice was criticised because it was considered that these requirements do not comply with the LOSC, and Finland and Sweden amended thereafter their legislation.[57] The only state having a sizeable navy, which requiring prior authorisation is China; the other states are coastal states with small navies.

Among the Nordic countries, Denmark has stipulated restrictions relating to the passage of foreign warships through the Danish Straits. The relevant Danish legislation (the Ordinance Governing the Admission of Foreign Warships and Military Aircraft to Danish Territory in Time of Peace, of 16 April 1999) regulates the approach to the Baltic Sea. Simultaneous passage of more than three warships of the same nationality through the Great Belt, Samsoe Belt or the Sound is subject to prior notification through diplomatic

53 "Uniform Interpretation of Norms of International Law Governing Innocent Passage" *Law of the Sea Bulletin* 12, No. 14 (1989).

54 Art. 25 UNCLOS.

55 Art. 18, 19 and 25.

56 See Roach and Smith (2012) 250–1.

57 Hakapää, *Uusi kansainvälinen oikeus* (2010) 394; Government Proposal HE 12/1996 vp.

channels.[58] Denmark defines the Danish Straits as straits governed either in whole or in part by long-standing international conventions in force, in the same way as Finland and Sweden have defined the Åland Strait. While signing and ratifying the LOSC, Finland and Sweden issued a Declaration that Article 35(c) covers the 1921 Åland Convention as an exception to the right of transit passage (see below section 3.3.2).

3.3 Navigational Rights in the Territorial Waters of the Åland Islands

The way the above general rules of the law of the sea affect more immediately the Åland Islands will be developed next.

3.3.1 General Issues Relating to the Territorial Waters of the Åland Islands

When the 1921 Åland Convention was signed, the Åland Islands were considered to be a delimited area under Finnish sovereignty, whose demilitarised and neutralised status was secured by parties. The obligations and limitations established within this convention have been written in such a way as to oblige all parties to act accordingly, but they allow Finland some exceptional arrangements. These exceptions widen the scope of Finnish jurisdiction in the area and are linked to Finnish sovereignty.[59] The convention also includes the right of innocent passage for foreign warships when navigating territorial waters, which reflects a will to preserve the freedom of navigation in the area. The object and purpose of the 1921 Åland Convention was to ensure that a military presence within the area would not become a threat to coastal states around the Baltic Sea. If not more than the land areas had been considered to be part of the demilitarised and neutralised area, the sea area surrounding the Åland Islands would have been part of Finnish territorial waters without any special status. Had this been the case, Finland would have had a similar right to naval operations within the archipelago as it would have enjoyed within other parts of its territorial sea and internal waters. Such development was rejected by defining the sea area around the archipelago as a demilitarised and neutralised zone.

3.3.2 Navigation in the Territorial Waters of the Åland Islands

The 1921 Åland Convention sets up limitations concerning navigation within the territorial waters of the Åland Islands. Although it is part of the exercise

58 Ordinance Governing the Admission of Foreign Warships and Military Aircraft to Danish Territory in Time of Peace, 16 April 1999

59 R. Erich, *Suomen valtio-oikeus I osa* (1924) 181–2.

of state sovereignty for states to grant preferential treatment to their own domestic vessels whenever passage through territorial waters is concerned, this approach is not possible for Finland within parts of its internal waters and territorial sea falling within the scope of the 1921 Åland Convention.[60] Bearing in mind the understanding of the limitations imposed by the 1921 Åland Convention, it is clear that Finnish sovereignty is limited in those parts of the Åland Islands' territorial waters that are regarded as internal waters and territorial sea, when compared to other parts of Finland's territorial waters. Furthermore, the 1921 Åland Convention blocks Finland's right to engage in bilateral or multilateral agreements, which could, for instance, implicitly or explicitly allow the passage of foreign warships through the sea area situated within the demilitarised and neutralised zone.

Defence material exports in transit through Finnish territory into a third country are, in general, regulated by Finnish law and are transportable only if authorised.[61] This kind of transportation and transit is not permitted in the demilitarised and neutralised zone on account of Article 4(1) of the 1921 Åland Convention.[62] According to Article 4(2) (b), during peacetime, the Finnish Government may grant only one warship belonging to a foreign state the right to enter the archipelago and temporarily anchor there. Consequently, a foreign military force's sojourn is not completely forbidden according to the convention. With regard to the prohibition of military forces and arms transportation, the 1921 Åland Convention does not make a distinction between merchant ships and warships. One must, nevertheless, also bear in mind that the right of innocent passage is beyond the scope of this regulation.

During wartime, the zone is considered to be a neutral area and should not be used for any activities linked to military operations. However, an exception to this rule exists in relation to the territorial waters of the Åland Islands because Finland has a right to lay mines in these waters and should be able to engage in such precautionary measures in order to preserve the security of the region.

In a similar way, Finland is permitted to send its armed forces into the zone during peacetime and keep them there temporarily provided this is strictly necessary to maintain order. One or two Finnish light surface warships have a right to visit the islands from time to time and to anchor there temporarily. Under special circumstances, Finland also has the right to bring other surface warships into the zone to a total limit of 6 000 tons.[63] The Finnish Navy's right to enter and stay in the Åland Islands' sea area was an

60 *Yearbook of the International Law Commission*, Volume I (1954), 122.
61 Act on the Export of Defence Materiel, Statutes of Finland 282/2012.
62 Section 1(1) of the Act.
63 It can be noted that modern warships of the Finnish Navy are often not more than 300 tons.

issue during negotiations leading up to the 1921 Åland Convention. Finnish and Swedish views seemed to oppose each other and Finland sought an unlimited right to send warships to the Åland Islands.[64]

During the negotiations, a number of states felt that the narrowness of the Gulf of Bothnia was worth taking into account, especially when considering whether it was appropriate to expand the zone's sea area more than necessary. States were particularly concerned about technical issues relating to the western side of the archipelago because there would be no passage route beyond these sea areas. They also highlighted the importance of maintaining sea traffic fluency around the archipelago.[65] Negotiations led to a compromise. Finland did not gain an unlimited right to send warships into the zone, and Sweden's efforts were likewise unsuccessful in so far as they aimed to limit the presence in the zone of warships belonging to the power holding sovereignty over the area. It is also worth mentioning that Sweden's proposal had not mentioned Finland and only spoke of a 'sovereign power'. If the conference had accepted Sweden's proposal to limit sojourn of warships in the zone, it would have had an influence on the other parties' potential to enter the zone. It seems obvious that this was also a reason why Article 5 contains the right of innocent passage through the territorial waters of the Åland Islands. The French delegation alluded to the fact that the sovereignty of Finland had to be respected because, according to international law, a state has the right to forbid foreign warships from entering its waters. Furthermore, the same delegation added that some coastal states already required prior notification from foreign warships before entering into their territorial waters. The delegation of Germany stated that the text of the convention was in accordance with international law because it only limits the authority of Finland but does not affect the general rule of international law regarding access. Finland stated that it must have a right to prohibit entry of foreign warships to its territorial waters.[66]

The 1921 Åland Convention does not identify exceptional circumstances. However, it does allow armed forces to enter the zone in order to maintain peace and stability in the region. Finland is allowed to grant a right of access to the zone (including a right to temporarily anchor there) to one warship of

64 "Conférence Relative à la Non-Fortification et à la Neutralisation des Iles d'Aland," *Actes de la Conferénce*, 21, 55–64; N. Fagerlund, *Innebörden av uttrycket temporärt förankra i de till öarna hörande vatten i artikel 4 st. 2b) Ålandskonventionen* (1994) 7. See also section 2.2 of the present book.

65 "Conférence Relative à la Non-Fortification et à la Neutralisation des Iles d'Aland," *Actes de la Conferénce*, 31.

66 "Conférence Relative à la Non-Fortification et à la Neutralisation des Iles d'Aland," *Actes de la Conferénce*, 21, 55–64; Fagerlund (1994) 7.

another state at a time. However, a question worth asking is why the other parties accepted this provision. Perhaps the reason for this was that they wanted to secure the possibility to visit the Åland Islands for themselves, which also meant putting additional pressure on Finland so that it complies with the convention. Nevertheless, it is noticeable that the convention does not oblige Finland to permit access to its territorial waters. It is also noteworthy that restrictions to access the demilitarised zone cover merchant ships and other non-military ships if they transport foreign or Finnish troops or arms and other implements of war.[67] Hence, the presence of Finnish and foreign warships in the demilitarised and neutralised zone is regulated by the 1921 Åland Convention. For instance, any anchoring of warships taking place in the zone can only be allowed temporarily. In practice, Finland has interpreted this requirement in such a way that warships of the Finnish Navy do not sojourn in the zone for more than 48 hours. The Finnish Navy informs the authorities of the Åland Islands about its visits to the zone. It is noteworthy that Finland is not obliged to adhere to these arrangements but does so voluntarily. Finnish practice indicates a willingness to undertake activities that are mentioned in the convention out of good faith and in co-operation with the local authorities based on the Islands.[68] It is also noteworthy that Article 4 refers to 'surface warships', which means the exclusion of Finland's submarines. To the extent that Finland has submarines, it has no right to bring them to the zone.[69]

The right of access to internal and territorial waters is left to Finnish jurisdiction and for this reason the 1921 Åland Convention does not oblige Finland to grant it.[70] This arrangement might also be said to illustrate the will of states to create a separate legal status for the Åland Islands. The starting point for many analyses concerning restrictions imposed by the 1921 Åland Convention tends to focus mostly on Finland's role in regulating passage of foreign warships.[71] Even though attention should be duly given to Finland's role when interpreting the convention, a more conservative approach would suggest that Finland enjoys full sovereignty up until the point where treaty arrangements restrict it specifically.[72]

67 Speech by the Minister of Foreign Affairs Erkki Tuomioja at a seminar, "*Åland och demilitarisering i dag*," on 7 March 2005 in Mariehamn.

68 Finnish Parliament Documents, *Kirjallinen kysymys* (written question) 457/2003 vp—Jaakko Laakso, Roger Jansson; Finnish Ministry for Foreign Affairs' response to Finnish Parliament speaker, regarding written question SS 457/2003, 12 November 2003.

69 Rotkirch (1986) 369.

70 See the case of the Russian school ship *Kruzenstern* at the beginning of this chapter; HBL 28.8.2017 and several articles in Finnish and Ålandic news outlets.

71 "Conférence Relative à la Non-Fortification et à la Neutralisation des Iles d'Aland," *Actes de la Conférence*, 63; Björkholm and Rosas (1990) 71; Rosas (1997) 33.

72 Björkholm and Rosas (1990) 71; Rosas (1997) 33.

As mentioned previously, the only exception to the coastal states' sovereignty over their territorial seas is the right of innocent passage of foreign ships. According to Article 5 of the 1921 Åland Convention, warships enjoy a right of innocent passage through the territorial waters of the Åland Islands. This same article does not mention the requirement of prior authorisation or notification for the innocent passage of warships; it only refers to the rules of international law in force. Prior notification of passage of warships emerged as a subject of debate during the 1921 conference negotiations. However, there was no mention of this in the convention's text. As mentioned already, the French delegate pointed out that the introduction of such prior notification regulations would not have brought about new forms of practice, as it was already commonly applied.[73] This indicates that other parties to the convention had also not opposed the position held by Finland at the time, which required prior notification before navigating Finnish territorial seas. In summary, it seems that other parties were willing to tolerate minor limitations to the passage of warships because such restrictions reflected the rules of international law and practice. At the time of the 1921 conference, coastal states could indeed require prior notification.

Innocent passage may not be prohibited. The right of innocent passage is a fundamental right of the law of the sea, and for this reason it can be only suspended temporarily. Up until the time when Finland changed its legislation on this issue, it had required prior notification from warships proceeding through its territorial waters. This legal evidence indicates how Finland had, in fact, accepted the principle of a right of innocent passage for warships. Article 5 of the 1921 Åland Convention only refers to existing international law and practice. Thus, at least with regard to the Åland Convention, Finland's stance on prior notification was justified in terms of existing international rules and practice.

As discussed previously, the delimitation in the Märket Reef posed particular challenges for Finland and Sweden. Finland made the following declaration upon signature and confirmed upon ratification of the LOSC:

> It is the understanding of the Government of Finland that the exception from the transit passage regime in straits provided for in Article 35 (c) of the Convention is applicable to the strait between Finland (the Åland Islands) and Sweden. Since in that strait the passage is regulated in part by a long-standing international convention in force, the present legal régime in that strait will remain unchanged after the entry into force of the Convention.

73 "Conférence Relative à la Non-Fortification et à la Neutralisation des Iles d'Aland," *Actes de la Conférence*, 63.

Sweden gave a corresponding declaration.[74] The parallel declarations issued by Finland and Sweden state that it is sufficient that part of the strait is governed by a long-standing treaty already in force. Consequently, the LOSC provisions on transit passage do not affect the legal regime of the Åland Strait where the regime of non-suspendable innocent passage is thus applicable. Other states did not question this position during the conference—not even the United States. However, the US Department of State has expressed in its publication "Limits in the Seas" that Article 35 (c) is not applicable to the Åland Strait because the territorial sea of the Åland Islands only extends by three nautical miles from the low-water line.[75]

All in all, the demilitarisation sets restrictions to the entry and presence in the sea area around the Åland Islands. The 1921 Åland Convention regulates access to the zone and Finland's jurisdiction over the zone. It also contains regulations for Finnish warships and foreign warships. The convention only allows Finnish light surface vessels to navigate the Åland Islands sea area with certain restrictions applicable during peacetime, but it does not mention submarines; it seems that the 1921 Convention equates Finnish submarines with foreign submarines. As far as is known, the number of visits of foreign warships and government ships in the demilitarised zone is low. It seems that Finland has taken the overall position to keep foreign warships away from the zone if their presence is not appropriate. The only exception to the coastal state's sovereignty over its territorial waters is the right of innocent passage and, as discussed above, the 1921 Åland Convention takes this old rule of the law of the sea into account in Article 5.

One of the current challenges in the exercise of innocent passage is its relationship to the Proliferation Security Initiative (PSI), which represents a new mode of co-operation. The PSI is a non-treaty-based activity established in 2003 upon a US initiative. It is a network of countries committed to the non-proliferation of weapons of mass destruction and has focused primarily on ocean transport. Finland has been a participant since 2004. The PSI targets merchant ships and participants have adopted the Statement of Interdiction Principles (SOP). Participants are committed to taking appropriate actions to stop and/or search suspected vessels in their internal waters, territorial seas or contiguous zones. The SOP includes elements which may contradict with the LOSC, including by requesting states participating in the PSI to stop and/or search, in their territorial waters, vessels "reasonably suspected" of carrying cargoes to or from states or non-states actors

74 See Declarations or Statements upon UNCLOS Ratification.

75 US Department of State, *Limits in the Seas No. 112*, United States Responses to Excessive Maritime Claims (1992) 66–7: "The United States, which is not a party to this Convention, has never recognized this international strait as falling within the Article 35(c) exception".

"of proliferation concern".[76] Interdiction operations against foreign vessels in the demilitarised zone would most likely be carried out by the Border Guard, rather than Finnish military authorities. Finally, interdiction operations might be problematic in the demilitarised zone of the Åland Islands even against merchant ships flying the Finnish flag if executive assistance is requested from military authorities.[77]

76 See the Interdiction Principles.

77 See P. Kleemola-Juntunen, "The Right of Innocent Passage: The Challenge of the Proliferation Security Initiative and the Implications for the Territorial Waters of the Åland Islands," in *The Future of the Law of the Sea: Bridging Gaps Between National, Individual and Common Interests*, ed. G. Andreone (2017) 239–69.

4 Regional Security Co-operation and the Åland Islands

After reviewing core concepts and questions, followed by an overview of the historical and legal background and content of the demilitarisation and neutralisation of the archipelago, we now turn to present-day forms of military and security co-operation in the Baltic Sea. We show the expansion and multiplicity of such co-operation, the increased fusion of military and civil aspects as well as the difficulties in ensuring adequate democratic control in such activities.

As a region situated in the middle of the Baltic Sea, as a border region and as part of the Republic of Finland, Åland is involved in regional co-operation both in the Baltic Sea region and in Europe as a whole. This chapter looks at such co-operation from a variety of perspectives relevant for the demilitarised Åland Islands. It starts with conceptual issues related to security co-operation, briefly introduces security integration in the EU and moves on to regional co-operation. Subsequently, the authors present one example of bilateral, Nordic co-operation, SUCBAS (Sea Surveillance Co-operation Baltic Sea). The period of the last twenty years, witnessing Finland's enhanced co-operation with NATO as well as shifts in Finnish views on multilateral and regional security, will also be discussed. In the present chapter, the Åland Islands are looked at in the context of regional co-operation and integration. The demilitarisation of the archipelago was intended to prevent war and decrease tension in the Baltic Sea and Northern Europe. Regional co-operation may also have the same purpose, as is shown in this chapter. Furthermore, regional co-operation can serve to safeguard the autonomous and demilitarised status of the islands, if pursued with such an intention.

Although the demilitarisation of the islands can be considered as a multilateral arrangement, it may simultaneously be part of regionalisation in the Baltic Sea region. Indeed, the 1921 Convention covers all Baltic Rim states, except for Lithuania, which did not have a Baltic coast in 1921, and Russia, which was not yet recognised and was not a member of the League of Nations. In the 1990s, the Baltic Sea seemed less tense after the end of the Cold War. For instance, Russia, Finland and Sweden joined the NATO Partnership for Peace programme in 1994 and participated regularly in the annual Baltic Operations (BALTOPS) exercise in the Baltic Sea. Russia has not, however, participated in the exercise since 2012 and declined to participate in SUCBAS.[1]

Much of the co-operation in the region, however, excluded security matters from the agenda and chose to focus on issues of environment or mobility, as was the case in the Helsinki Commission (HELCOM) and in the Council of the Baltic Sea States (CBSS). The CBSS started operating in 1992 and includes eleven states: Finland, Sweden, Norway, Iceland, Russia, Estonia, Lithuania, Latvia, Poland, Belarus and Germany. Foreign ministers and a representative of the European Commission are supposed to meet regularly, but government-level meetings have not been held since June 2014 as a result of EU sanctions against Russia, which all the other CBSS countries adopted. However, the Committee of Senior Officials still functions normally in the council and the organisation is in operation.

4.1 Finland, Åland and the European Union (EU)

Turning to current security co-operation, the most important security organisation for Finland is undoubtedly the EU, and the most important reason for joining the EU was security, as argued by the Finnish President at the time, Mauno Koivisto.[2] After the end of the Cold War, Finland became a member of the EU, committing to the Common Foreign and Security Policy (CFSP) of the Union established in the Maastricht Treaty (1992). The EU has increased its security co-operation since the 1990s. Finland became a member state of the EU in 1995, together with two other non-NATO countries, Sweden and Austria.

The Åland Islands were addressed in the accession negotiations, mainly with regard to the exemptions granted for the islands related to taxation and the right of domicile. The only implication to the demilitarised and neutralised status of the Åland Islands can be read in the preamble of the Protocol No. 2 on the Åland Islands attached to the Accession Treaty: "[t]aking into account the special status that the Åland Islands enjoy under international law". Interestingly enough, it was the commission that proposed mentioning the demilitarised status, but only a general reference to the international

1 B. Österlund, "Changing Scenarios in the Baltic Security Policy From the Historical Perspective," *Journal of East-West Business* 19, No. 1–2 (2013) 74.

2 M. Koivisto, *Witness to History: The Memoirs of Mauno Koivisto, President of Finland 1982–1994* (1997) 246.

legal status of the archipelago remained in the final version.[3] The Finnish Government Bill 135/1994 confirms this decision, arguing that the intention of the entry was to justify the exemptions granted for Åland and only relates to linguistic and cultural rights guarantees in the League of Nations Convention of 1921. The Finnish Parliament Constitutional Committee, however, stated that the entry seems more extensive than just relating to linguistic and cultural rights.[4] Overall, the demilitarised status of the Åland Islands did not receive much attention in the accession process,[5] but the focus was on exemptions established for the islands.

After EU accession, the demilitarised status of the Åland Islands was also confirmed in subsequent EU documents. In the Lisbon Treaty (2007) amending the basic treaties of the EU, it was stated that the Åland Protocol would continue to apply.[6] Already in the treaty negotiations in the early 2000s, Finnish Prime Minister Matti Vanhanen specified that maintaining the position of the islands was one of the Finnish objectives that had been reached.[7] In 2010, following a request by Ålandic politicians, Finland gave a unilateral declaration at a meeting of the permanent representatives (COREPER), stating that the demilitarised and neutralised status of the islands remains in force even after the entry into force of the Lisbon Treaty.[8] Due to the autonomy of the Åland Islands, Ålandic representatives can actively participate in the EU affairs of Finland, which often also concern the Åland Islands.[9] In addition, a separate referendum on accession to the EU was organised in the islands and the Ålandic politicians were able to decide whether to approve the Lisbon Treaty that entered into force in 2009.

As a non-NATO country, Finland has had an ambiguous approach to the development of the security and defence policy within the Union. When the country joined the EU, it became only an observer in the defence alliance attached to the EU, the WEU. Finland had reformulated its previous

3 N. Fagerlund, "The Special Status of Åland Islands in the European Union," in *Autonomy and Demilitarisation in International Law: The Åland Islands in Changing Europe*, eds. L. Hannikainen and F. Horn (1997) 196.

4 Finnish Parliament Constitutional Committee, "Statement of the Committee on Constitutional Law on Government Bill 135/1994," 1994.

5 Y. Poullie, *Åland's Demilitarisation and Neutralisation at the End of the Cold War* (2016).

6 "Consolidated Versions of the Treaty on European Union and the Treaty on the Functioning of the European Union," *Official Journal of the European Union* (2016) Article 355(4). "Treaty of Lisbon," *OJ* C 306 (2007).

7 M. Vanhanen, "*Pääministerin ilmoitus eduskunnalle EU:n hallitustenvälisen konferenssin johdosta* [Prime Minister's Announcement Made to the Parliament Concerning the Intergovernmental Conference of the EU]," 22 June 2004.

8 Council of the European Union, "2298th Meeting of the Permanent Representatives Committee Held in Brussels on 2–4 and 7 December 2009," Brussels, 2010.

9 Chapter 9a of the Act on the Autonomy of Åland (1991/1144).

neutrality policy as military non-alliance, and did not want the WEU to be integrated into the Union. In order to prevent this, Sweden and Finland proposed that only the crisis management tasks of the WEU be integrated into the Union, which took place in 1999. Preventing full integration may not have been the only reason; for example, Finland and Sweden also received more influence in the WEU. The CFSP was complemented with the European Security and Defence Policy (ESDP) that was launched the same year, though mainly dealing with crisis management. This was not the end of the defence integration process, but WEU became *de facto* integrated into the Union in the 21st century, when member states started to prepare the Constitutional Treaty of the EU. Although the envisioned Constitutional Treaty did not become a reality, the Lisbon Treaty, in force since 2009, contained the same defence clauses that were introduced in the constitutional draft, including those concerning permanent structured co-operation, mutual assistance and solidarity.

In the EU treaty negotiations concerning the assistance clauses, the militarily non-allied countries (Finland, Sweden, Austria, Ireland) convened and proposed that the mutual assistance clause be reformulated in the form that member states "may request" aid rather than states being obligated to provide that.[10] They issued a letter to the President of the Council of the European Union stating that 'formal binding' would not be compatible with the security policies of these countries.[11] Eventually, the mutual assistance clause (42(7) of the Treaty on European Union) was complemented with a provision that "[t]his shall not prejudice the specific character of the security and defence policy of certain Member States". The problematic was raised in particular by the Irish politicians who needed assurances of maintaining their policy of neutrality,[12] and a Protocol on the concerns of the Irish people on the Treaty of Lisbon was also appended to the treaty. The Protocol stated that the (renamed) Common Foreign and Security Policy (CFSP) "does not prejudice the security and defence policy of each Member State, including Ireland, or the obligations of any Member State".[13] In 2007, the year when the Lisbon Treaty was signed, Finland reformulated its military non-alignment stance as 'non-membership of a military alliance', which should pose no problems for further defence co-operation within the EU.

10 H. Ojanen, "Finland: Rediscovering Its Nordic Neighbours After an EU Honeymoon?," *Security Dialogue* 36, No. 3 (2005) 410.

11 Conference of the Representatives of the Governments of the Member States, "Cover Note CIG 62/03," 5 December 2003.

12 K. Devine, "Neutrality and the Development of the European Union's Common Security and Defence Policy: Compatible or Competing?," *Cooperation and Conflict* 46, No. 3 (2011) 354.

13 "Protocol on the Concerns of the Irish People on the Treaty of Lisbon," *Official Journal of the European Union*, 2 March 2013.

With the Lisbon Treaty, the security guarantees of the WEU were incorporated in the so-called mutual assistance clause: "[i]f a Member State is the victim of armed aggression on its territory, the other Member States shall have towards it an obligation of aid and assistance by all the means in their power" (Article 42(7) TEU). Another EU assistance provision is the so-called solidarity clause (Article 222 TFEU), which obliges member states to provide assistance in case of a terrorist attack or a natural or man-made disaster. So far, only the mutual assistance clause has been activated (following the terrorist attacks in Paris in November 2015). At that time, Finland also provided help to France in crisis management operations, although Act 418/2017 on decision-making concerning the provision and reception of international assistance entered into force only in July 2017. The act was passed after amendments that involve the parliament in the decision-making even in urgent matters; in urgent situations involving military measures, the government or the president may make the decision after providing a report for the parliamentary Committee for Foreign Affairs. Thereafter, they must immediately bring the matter to the parliament, which decides on the continuation and termination of the assistance.

One way to deepen practical defence co-operation is provided by the Permanent Structured Cooperation (PESCO) introduced in Articles 42(6) and 46 TEU, which would be limited to a group of member states voluntarily harmonising their defence together. If carried out in the near future, PESCO can be expected to include at least Germany and France, which issued a joint demand for more defence co-operation, even though their conceptualisations seem to vary, with Germany prioritising wide participation and France pushing for hard military criteria. The 2006 German White Paper on Security Policy explicitly mentioned PESCO.[14] Finland has recently announced its willingness to participate in such co-operation in the future.[15] In November 2016, EU member states jointly called for the launch of the permanent structured co-operation,[16] and the European Parliament proposed adopting the PESCO and implementing the 2016 Global Strategy for the European Union's Foreign and Security Policy.[17] Ahead of the June 2017 European Council meeting, Finland underlined the need

14 J-M. Ayrault and F-W. Steinmeier, "A Strong Europe in a World of Uncertainties" (2016) 1–4; Federal Government of Germany, "White Paper on German Security Policy and the Future of the Bundeswehr," 13 July 2016.

15 Finnish Parliament, "Minutes of the Plenary Session on 8 March 2017," 2017.

16 Council of the European Union, "Implementation Plan on Security and Defence," 14 November 2016.

17 European Parliament, "European Parliament Resolution of 22 November 2016 on the European Defence Union (2016/2052(INI))," 22 November 2016.

to launch PESCO,[18] repeated also by the European Council.[19] Finland's strong engagement in this process is also illustrated in the establishment of a new institution, the European Centre of Excellence for Countering Hybrid Threats. The Memorandum of Understanding for this centre was signed in April 2017 and it became operational in October 2017; its institutional status and legal regime deserves further scientific analysis in the future.[20]

With regard to the demilitarised Åland Islands, it has also been envisioned that Finland could appeal against the new provisions in case the Ålandic status was compromised.[21] Indeed, all the signatories to the 1921 Convention are currently EU member states (though the UK is about to leave the EU), which would make it reasonable to appeal to the EU, too, rather than simply to the signatories of the convention. Russia might oppose such a decision, since the 1921 Convention (of which Russia is not a party) only provides the possibility to appeal to the League of Nations and to the signatories.

Some politicians have argued that deeper defence co-operation could mean that the EU is becoming militarised (see also the Introduction to this book). In addition to political discourses, militarisation has indeed been used in scholarly writings with regard to the EU. A number of scholars argue that EU policies such as those on normative power, cyber security, police or diplomacy are becoming militarised within the EU.[22] For example, Manners suggests that militarisation decreases the normative power of the EU and that the 2003 European Security Strategy served as a symbolic signpost of militarisation.[23] The development of the EU defence policy does not, however, necessarily have to be seen in terms of militarisation. Other scholars maintain that defence integration is as a logical continuation of an "ever closer union", where exogenous factors such as the Yugoslavia wars after

18 Finnish Government, "*EU-ministerivaliokunnan kokous linjasi kantoja Eurooppa-neuvostoon* [Meeting of the Ministerial Committee on European Union Affairs Formulated Stances for European Council Meeting]," 20 June 2017.

19 European Council, "European Council Meeting (22 and 23 June 2017)—Conclusions," 23 June 2017.

20 See European Centre of Excellence for Countering Hybrid Threats (Hybrid CoE).

21 H. Lax, "The Åland Regime After Crimea," *Baltic Rim Economies Review*, No. 5 (2015) 10.

22 I. Manners, "Normative Power Europe Reconsidered: Beyond the Crossroads," *Journal of European Public Policy* 13, No. 2 (2006) 182–99; C.M. Constantinou and S.O. Opondo, "Engaging the 'Ungoverned': The Merging of Diplomacy, Defence and Development," *Cooperation and Conflict* 51, No. 3 (2016) 307–24; K. Friis and E. Rechborn-Kjennerud, "From Cyber Threats to Cyber Risks," in *Conflict in Cyber Space: Theoretical, Strategic and Legal Perspectives*, eds. K. Friis and J. Ringsmose (2016) 27–44; T. Weiss, "The Blurring Border Between the Police and the Military: A Debate Without Foundations," *Cooperation and Conflict* 46, No. 3 (2011) 396–405.

23 Manners (2006).

the Cold War have contributed.[24] From a functional perspective, Biscop has even considered that it is crucial for European states to co-operate in order to maintain their military capability.[25] Such concerns may lie behind the Finnish promotion of deeper European defence co-operation, coupled with perceptions of a tense security situation, which the government has described in its recent white papers.[26]

Despite the tendency towards more security co-operation, the EU is struggling in the conflict of pressures between hard and soft means, as well as supranationalism and intergovernmentalism. The EU used to be considered a soft power under different names. In the 1970s, it was characterised as a civilian power,[27] and in the 21st century as a normative power[28] or as an ethical power.[29] In addition to the traditional image of a promoter of peace, there are also other major bottlenecks in establishing a common defence policy, despite support from different angles. One relates to different geographical priorities with countries to the east emphasising the Eastern neighbourhood, Mediterranean countries prioritising the Middle East and North Africa and so on. Furthermore, there are differences in strategic cultures, attitudes towards the US and the quality of armed forces.[30] Other obstacles include military inflexibility, legal, conceptual and practical problems in co-ordinating civil and military activities, and financial constraints.[31] Even with some political determination, the EU is still far from having a common defence.

The demilitarisation of the Åland Islands is a *sui generis* arrangement, which is why it can be difficult to locate in the European security context. It is a long-standing arrangement, but it simultaneously seems to represent a modern solution in the face of increasing regional juxtapositions. It shifts attention from the military to the diplomatic sphere and from the national

24 J. Howorth, "Why ESDP Is Necessary and Beneficial for the Alliance," in *Defending Europe: The EU, NATO, and the Quest for European Autonomy*, eds. J. Howorth and J.T.S. Keeler (2003) 221–2.

25 S. Biscop, "All or Nothing? The EU Global Strategy and Defence Policy After the Brexit," *Contemporary Security Policy* 37, No. 3 (2016) 431–45.

26 Prime Minister's Office Finland, "Government Report on Finnish Foreign and Security Policy 6/2016" (2016); Prime Minister's Office Finland, "Government's Defence Report VNS 3/2017," *Government Report 5/2017* (2017).

27 F. Duchêne, "Europe's Role in World Peace," in *Europe Tomorrow: Sixteen Europeans Look Ahead*, ed. R. Mayne (1972) 32–7.

28 Manners (2002) 235–58.

29 L. Aggestam, "Introduction: Ethical Power Europe?," *International Affairs* 84, No. 1 (2008) 1–11.

30 A. Bakker et al., "Spearheading European Defence: Employing the Lisbon Treaty for a Stronger CSDP," Netherlands Institute of International Relations, *Clingendael Report* (2016).

31 On civil-military co-operation see section 1.3 above.

to the international sphere. Whereas Finland's foreign policy stance appears to rest on a rather flexible concept in the EU, demilitarisation is strictly binding for all parties. Notwithstanding issues of legitimate expectations and unilateral acts under international law, foreign policy stances are, even if provided in the constitution or imposed in a peace treaty—such as for Austria after the Second World War—a self-declaration of a state which does not create any immediate obligations for other states. Demilitarisation and neutralisation, in turn, can even be regarded as an objective regime or as customary international law, in which case the agreements would bind everyone, not just parties to the agreements.[32] The EU has also acknowledged the international legal status of the islands, but the demilitarisation of the Åland Islands is not specifically mentioned in EU documents concerning security co-operation.

The EU is currently facing a difficult challenge in its attempts to enhance military co-operation, in which all member states may not be willing to participate. Although the UK, traditionally the fiercest opponent to EU defence integration, is about to leave the EU, the approach of the non-NATO countries (Finland, Sweden, Austria, Ireland, Cyprus and Malta) towards defence co-operation varies. Sweden, Ireland and Austria, for example, seem more hesitant than Finland when it comes to defence integration, although this does not necessarily mean that they would remain outside PESCO. It is possible that Brexit will lead to a multi-speed Europe in defence, with only willing member states participating in the permanent structured co-operation or a similar arrangement. Having military capability would certainly be a considerable change in the integration process of the European Union, which so far has focused mainly on diplomacy in its external relations. For example, diplomacy in terms of sanctions was the first response to the Russian annexation of Crimea. In 2014, the member states also agreed not to have regular bilateral summits with Russia.[33] This also led to the freezing of high-level meetings in the Baltic Sea region, which will be discussed in section 4.2 below.

In a similar vein to the Finnish Government, Ålandic politicians seem to have a positive approach towards European co-operation, even as regards defence. This became clear when the Government of Åland drafted Guidelines for the Government Opinion on Åland's demilitarisation and neutralisation in 2013. In this document, the Government of Åland assumed that demilitarisation and neutralisation was not an obstacle to the development of

32 T. Tiilikainen, "Åland in European Security Policy," in *The Nordic Countries and the European Security and Defence Policy*, eds. A.J.K. Bailes et al. (2006) 349–55.

33 European Council, "European Council Meeting (20–21 March 2014)—Conclusions," 21 March 2014.

EU foreign and security policy. They also considered that the EU strengthens the autonomous and the demilitarised status of the islands through reconfirmations in several documents, including in the basic treaties of the Union.[34]

4.2 Integration in the Baltic Sea Region and the Åland Islands

Since 2004, all the Baltic Sea coastal states with the exception of Russia are part of the EU, and the Baltic Sea region has also been addressed in EU policies. Co-operation and regional integration around the Baltic Sea is difficult to describe in a condensed way, partly due to the institutional heterogeneity, the scope of different priorities among countries and the EU's complicated approaches in dealing with the region.[35] Already in 1997, Finland proposed establishing a Northern Dimension policy for the EU to enhance co-operation in Northern Europe. Its purpose was to boost co-operation between the EU, Russia, Norway and Iceland. Interestingly enough, Sweden had proposed a Baltic Sea Region Initiative already in 1996, but the proposal did not lead to concrete action. In contrast, the Finnish initiative was adopted in 1999 and renewed in 2006, although current co-operation is rather inactive.[36] Indeed, the Northern Dimension policy came to a halt in 2014 due to EU sanctions against Russia.[37] The examples of the Northern Dimension and the CBSS show that regional organisations have not been able to go beyond state-centred policies, but that interstate tensions still matter. This is so even though hard security matters are excluded from the table of these organisations. Perhaps it also reveals how weak the Northern Dimension was in the first place, with no separate EU funding or new structures.

When the Baltic States joined the EU in 2004, Sweden made a proposal to establish an informal group of Nordic and Baltic States to co-ordinate positions in the EU Council. This came to be known as NB6 co-operation, consisting of the three Nordic EU states (Finland, Sweden and Denmark) and the three Baltic States (Estonia, Latvia and Lithuania). Sometimes Germany and Poland are also invited (NB6 + 2), such as when drafting

34 Parliament of Åland, "*Plenum Den 4 December 2013 Kl. 14.00*" (2016) 1–19. See also the final version, Government of Åland, "*Policy för Ålands demilitarisering och neutralisering. Handbok för landskapets myndigheter*" (2015) 16–18.

35 G. Herolf, "Co-operation in the North—Multilateralism or Mess?," *Mercury E-Paper No. 7* (2010).

36 C. Gebhard, "Soft Competition: Finland, Sweden and the Northern Dimension of the European Union," *Scandinavian Political Studies* 36, No. 4 (2013) 365–90.

37 N. Götz, "The Case of the Baltic Sea Area: Spatial Politics & Fuzzy Regionalism," *Baltic Worlds*, No. 3 (2016) 61.

the European Union Strategy for the Baltic Sea Region (EUSBR), adopted in 2009. Alongside the NB6 co-operation in the EU, there is co-operation between the Nordic and Baltic organisations. For example, the Nordic Council and the Baltic Assembly co-operate under what is called the Nordic-Baltic 8 (NB8). This co-operation started in 1992, and annual meetings with the Prime Ministers and foreign ministers of these countries are organised. Furthermore, the NB8 co-operation also sometimes includes the UK or the four Visegrad countries (Poland, the Czech Republic, Slovakia and Hungary) in its activities.

The EUSBR is the most recent comprehensive co-operation network in the region, covering environmental issues, safety and economy. It works in co-operation with the other regional organisations; the Secretariat of the CBSS, for example, co-ordinates a priority policy area on safety (titled 'Secure'), horizontal actions related to neighbours, sustainable development and bioeconomy as well as serves as a Task Force for Communication of the EUSBR. The strategy did not introduce any new funds, institutions or legislation, but rather constitutes a network co-ordinating regional co-operation. It was the first macro-regional strategy in the EU and intended to serve as a model for other regions in Europe.

It is remarkable that the EU has understood the value of regionalisation, proposed at a time when the nature of the region seemed to have changed for the better; policies of dominance were no longer relevant and there was less East-West antagonism.[38] It seems that the EU no longer considered regionalisation as a threat to European integration, as was perhaps the case before.[39] Instead, regionalisation appeared as a manner to further European interests at the regional level. Joenniemi indeed argues that the strategy was also about the development of the EU as such, aimed at its many macro-regions.[40] The Baltic Sea region was thus seen as a model which could help in furthering the EU's objectives. The reason for the strategy was perhaps not to promote regionalisation as such, but rather to promote the EU's interests through regionalisation.

An interesting question in all this regional co-operation is the role of security. As argued by Browning and Joenniemi in 2004, the Nordic states could be described as constituting a 'security community', where co-operation has never been associated with preventing war between the countries. However, this is not entirely the case with co-operation in the Baltic Sea region, where

38 P. Joenniemi, "The EU Strategy for the Baltic Sea Region: A Catalyst for What?," *DIIS Brief* (August 2009).

39 C.S. Browning, "Towards a New Agenda? US, Russian and EU Policies in Northern Europe', in *NEBI Yearbook 2003*, eds. L. Hedegaard and B. Lindström (2003) 273–89.

40 Joenniemi (2009).

security concerns have also been important. Instead of hard security, however, so-called soft security has been more important and concerns related, for example, to the climate, have simultaneously served to desecuritise interstate relations.[41]

As previously discussed, hard security and geopolitical relations have also had an effect on Baltic Sea co-operation. Interestingly enough, the work of the Arctic Council has continued uninterrupted despite the tense relations between Russia and Western states, perhaps because the Arctic Council has completely excluded security matters from its work. This difference might also reveal something about the relevance of these organisations: whereas Arctic co-operation might be considered more important and less related to interstate tensions, Baltic Sea region co-operation appears to be something that the governments can set aside. Despite the various organisations promoting co-operation in the Baltic Sea region, actual co-operation is still very much dependent on interstate relations. It seems that Browning and Joenniemi were right in their 2004 predictions that the regional co-operation in the Baltic Sea region might also result in resecuritisation of the relations.[42] Indeed, the threat of the 'other' and more realist security agendas continue to cast a shadow on co-operation in the region.

Although the Åland Islands play a minor role in regional co-operative arrangements, they are active in Nordic co-operation, which has been vivid ever since the Nordic Council was established in 1952. A passport union among the five Nordic countries (Finland, Sweden, Norway, Denmark and Iceland) was established in the 1950s along with the Nordic job market and harmonised social security systems. It is precisely this co-operation that Joenniemi and Browning consider being a 'security community'. According to them, at least since the 1950s, Nordic co-operation has been based on a sense of togetherness and similarity, and security and defence do not appear central in the co-operation.[43]

Co-operation in the Baltic Sea region is mainly non-military, but some defence co-operation arrangements have been put in place. A Nordic Defence Cooperation (NORDEFCO) network was established between the Nordic countries in 2009 and co-operation has also sometimes included the Baltic States. NORDEFCO merged three previous co-operation structures: peace support education and training (NORDCAPS), armament co-operation (NORDAC) and trilateral defence co-operation (NORDSUP) between Finland, Norway and Sweden. A Memorandum of Understanding (MoU) was

41 C.S. Browning and P. Joenniemi, "Regionality Beyond Security? The Baltic Sea Region After Enlargement," *Cooperation and Conflict* 39, No. 3 (2004) 233–53.

42 Ibid.

43 Ibid.

signed among the five Nordic countries on 4 November 2009, replacing previous arrangements. The matter was not brought to the Finnish Parliament, but the implementation of a later agreement concerning co-operation in the defence materiel area was discussed in parliament and approved in Act 140/2017. None of the Memoranda of Understanding signed under the umbrella of NORDEFCO has been debated in parliament, but the government regularly provides reports on state agreements for the parliamentary Committee for Foreign Affairs. In the subsequent corresponding report after the approval of the MoU concerning NORDEFCO, the government stated that the MoUs are not legally binding, but parties do strictly comply with them. The report further states that in some cases, the agreements include provisions that fall under legislative affairs, in which case they need to be processed in parliament.[44] It seems that particularly in defence affairs, the government does not want to bring the issues under parliamentary debate unless strictly necessary. Neither was there any parliamentary debate on the Northern Group, established in 2010, a loose defence co-operation network that includes the UK, Netherlands and Poland in addition to the Nordic and Baltic States.

With regard to the demilitarisation of the Åland Islands, Finnish politicians, and even Defence Ministers, seem to have incomplete knowledge,[45] which is why parliamentary scrutiny becomes even more crucial. The Ålandic MP of the Finnish Parliament is obviously concerned over demilitarisation provisions, but MPs from other parties, such as the Left Alliance, have also presented written or oral questions concerning compliance with the demilitarisation provisions.[46] The government does not necessarily address the demilitarisation of the Åland Islands when discussing security and defence policy, but it is often the MPs that want to hear that the government is committed to the demilitarisation provisions (see also section 4.4 in this book). The parliamentary debates thus underline Finland's commitment to international law.

Among all the regional organisations discussed in this section, the Åland Islands have their own representatives only in the Nordic Council, whereas the Finnish Government represents Finland as a whole in the other organisations. The Nordic Council is an interparliamentary body with eighty-seven members, out of which two representatives come from the Åland Islands.

44 Finnish Government, "*Perustuslain 97 §:n mukainen selvitys eduskunnan ulkoasiainvaliokunnalle Suomen kansainvälisestä sopimuspolitiikasta ja sen kehityssuunnista*" (2011).

45 J. Niinistö, "*Ahvenanmaan asia*," *Jussi Niinistö's Blog*, 17 October 2016.

46 Finnish Parliament, "Written Question 457/2003 vp: Compliance With the Agreement Arrangements Concerning the Demilitarisation of the Åland Islands," 12 November 2003; Finnish Parliament, "Minutes of the Plenary Session on 20 Oct 2016: Oral Question on the Demilitarisation of the Åland Islands," 20 October 2016.

In the Nordic Council of Ministers, in turn, the Åland Islands have a representative but no separate voting power. All the self-governed areas of the Nordic Council (Åland Islands, Faroe Islands and Greenland) received more extensive representation in the Nordic Council after the adoption of the so-called Åland Document in 2007.[47] In 2017, Åland had a visible role in the Nordic Council, as the President of the Council was an Ålandic politician, Britt Lundberg. The other regional organisations discussed above function on a governmental basis, and the Ålandic decision-making bodies can influence regional matters mainly through the Finnish Government. Although the islands do not possess foreign policy power, they have the right to participate in decision-making about foreign affairs of relevance to the islands.[48] As regards the Baltic Sea, there are, of course, many such decisions.

In light of regionalisation, Åland could be one example of a society where the emphasis seems to be on sub-national and supranational governance instead of just being limited to sovereign states. Territorial autonomy and regional integration are both looked at as important.[49] In other words, Ålandic politicians appreciate the self-government of the islands, but support for Nordic co-operation is also strong.[50] The Nordic emphasis may also have to do with language, since the islands are Swedish-speaking, as opposed to the mainly Finnish-speaking mainland. In contrast, the official languages in other Nordic countries are all cognate languages to Swedish. Furthermore, the positive attitude may relate to the fact that the self-governed Nordic regions are recognised in Nordic co-operation. For example, in summer 2016, all the Nordic Prime Ministers gathered at a ministerial meeting at the Åland Islands, hosted by the Governor of Åland. Meetings are also held on the islands regularly. In addition to the co-operation fora discussed in this section, the next section provides a closer look at sea surveillance co-operation and its relevance for the demilitarised Åland Islands.

4.3 Bilateral and Nordic Co-operation: The Example of SUCBAS

On 4 March 2009, a seminar on co-operation in matters pertaining to territorial surveillance was held at the House of Estates in Helsinki. The participating states, six at the time, had been invited by Finland and Sweden,

47 S. Stephan, "Making Autonomies Matter: Sub-State Actor Accommodation in the Nordic Council and the Nordic Council of Ministers—An Analysis of the Institutional Framework for Accommodating the Faroe Islands, Greenland and Åland within 'Norden,'" *European Diversity and Autonomy Papers No. 3* (2014).

48 Act on the Autonomy of Åland (1991/1144) Chapter 9.

49 G. Prinsen and S. Blaise, "An Emerging 'Islandian' Sovereignty of Non-Self-Governing Islands," *International Journal* 72, No. 1 (2017) 56–78.

50 Parliament of Åland, "Minutes of the Plenary Session on 3 December 2013," 2013.

which had already been co-operating on matters of sea surveillance in the Baltic Sea within the so-called SUCFIS effort (Sea Surveillance Co-operation Finland–Sweden). The SUCFIS co-operation started as a bilateral exchange of information on naval exercises in 1999 and was in operation by 2006.[51] The six states signed a Letter of Intent in 2009.[52]

In October 2012, a so-called Technical Arrangement was concluded between the representatives of eight states around the Baltic Sea for the purpose of Sea Surveillance Co-operation [in the] Baltic Sea (SUCBAS).[53] The document was signed by various types of military actors from Denmark, Estonia, Finland, Germany, Latvia, Lithuania, Poland and Sweden (here in alphabetical order). Finland stands on top of the signing parties and participates in this co-operation through the Ministry of Defence, with Captain Dan Wilén, then Chief of Operations of the Finnish Navy, signing the document. The signature of Finland is followed by that of Sweden, the second initiator of the broadening of this co-operation. Other countries participated and signed either directly through the armed forces themselves, namely the navy (Sweden, Denmark, Estonia and Poland), or through the respective ministry of defence (Latvia, Lithuania and Germany). In 2015, the United Kingdom also entered this co-operation network.

In contrast to the treatment of the co-operation prompted by the Single European Sky directives of the EU, which created the North European Functional Airspace Block (NEFAB) also in year 2012, SUCBAS co-operation was not discussed by the Finnish Parliament.[54] In 2009, the Defence Committee of the Finnish Parliament referred to SUCBAS as a 'project' within the context of otherwise discussing EU maritime surveillance proposals.[55] In a bill presented by the Finnish Government to parliament in June 2016 there is brief mention of SUCBAS.[56] Under a heading titled "regional defence co-operation", and following a description of the Nordic Defence Cooperation (NORDEFCO), as well as of bilateral Finnish–Swedish defence co-operation,

51 C. Haglund, "The Baltic Sea as an Example of Regional Maritime Security Cooperation," *Baltic Rim Economies Review* No. 4 (2014) 4.

52 D. Åkerström, "*Nätverksbaserat försvar-Då och Nu,*" Försvarshögskolan (2012) 26.

53 See also the SUCBAS website.

54 S. Spiliopoulou Åkermark, "The Meaning of Airspace Sovereignty Today—A Case Study on Demilitarization and Functional Airspace Blocks," *Nordic Journal of International Law* 86, No. 1 (2017) 91–117.

55 Finnish Parliament Defence Committee Opinion 9/2009 concerning EU Commission proposal on the integration of maritime surveillance in COM(2009)538 final. It may be noted that in this Commission Proposal, SUCBAS is identified as a military co-operation (in endnote 10).

56 Finnish Government Bill RP 94/2016 on proposed changes of legislation pertaining to the Act on Armed Forces, the Territorial Surveillance Act and the Act on Conscription.

it is stated that co-operation in the field of territorial surveillance has been developing alongside deepened defence co-operation. For Finland, the most important co-operation of this kind in the maritime sphere is with MARSUR (Maritime Surveillance Networking), which has a European coverage, and SUCBAS. Those are "on their way to be operationalized", explained the government, even though such operationalisation seems to have been ongoing at least since 2009.[57] Interestingly, Swedish Navy Captain and Head of Operations Bengt Lundgren had, in an EU presentation in 2013 argued that both SUCFIS and SUCBAS were already fully operational at that time.[58]

SUCBAS is said to enhance capacity for oversight of territorial integrity at times of peace and it also enhances the possibility of creating reliable situational awareness in a situation of exception. As currently explained on the SUCBAS website, "information can be shared among national governmental institutions with a maritime responsibility regardless if these are civil or military".[59] While the concerns behind these initiatives are said to include maritime safety and environmental damage, it is clear by the composition of the national actors involved as well as by the operationalisation pursued that this is primarily a form of military and intelligence activity. Writing in October 2014, months after the occupation of Crimea by the Russian Federation, the Finnish Minister of Defence at the time, Carl Haglund, approached the issue of maritime safety first from the perspective of trade, and the fact that approximately half of Finnish trade is transported by sea, then from the perspective of environmental concerns and finally to wider security concerns.[60] Haglund ends by recalling that increased co-operation does not affect a state's "capacity and readiness to defend itself militarily", epitomising the Finnish view of "co-operation but independent capacity".

The fusion between civilian and military goals and actors is also visible in the Swedish Government enquiry made on the topic of maritime security in 2012.[61] The enquiry welcomed the establishment of SUCFIS and of SUCBAS and proposed both the expansion of SUCBAS to the North Atlantic (something which materialised when the United Kingdom became a participating state in 2015) and the progression and leadership by Sweden of EU efforts for a Common Information Sharing Environment (CISE).[62]

57 Ibid., 15.

58 B. Lundgren, "Security and Surveillance Cooperation in the Baltic," slides with notes.

59 See the SUBCAS website.

60 Haglund, "The Baltic Sea as an Example of Regional Maritime Security Cooperation" (2014).

61 Swedish Government, "*Statens Offentliga Utredningar, SOU 2012:48 'Maritim samverkan'*" (2012).

62 Ibid., 79–82. CISE is described as a "voluntary collaborative process".

In contrast to NEFAB co-operation and ratification, which followed the constitutional order in Finland, there was no parliamentary discussion and decision in the SUCBAS case. This means that issues pertaining to the demilitarisation and neutralisation of the Åland Islands, as well as other broader legal matters, including law of the sea and surveillance matters, were not discussed, at least publicly.

The legal nature of the SUCBAS Letter of Intent and later the Technical Arrangement is not entirely clear. The Technical Arrangement provides explicitly in its Section 8.2 that the document *does not constitute a treaty* under the Vienna Convention on the Law of Treaties. However, for matters pertaining to international affairs in Finland, the Finnish Constitution is at play. The 1999 Finnish Constitution entrenches not simply the principles of popular sovereignty and parliamentarism when providing that "Finland is a sovereign republic" and that "[t]he powers of the State in Finland are vested in the people, who are represented by the Parliament". It requires that the exercise of all public powers shall be based on a legal act and, furthermore, that the President of the Republic is the commander-in-chief of the Finnish defence forces as well as decides on matters of war and peace, thus ensuring the civilian control over the military (Sections 58, 93 and 128).[63]

Chapter 8 of the Finnish Constitution concerns decision-making in matters of international relations. Section 94(3) provides very succinctly that international obligations should not endanger the democratic foundations of the constitution. This is an obligation binding for all state authorities in Finland, including the government, the Navy and the Coast Guard, when engaging in international co-operation.[64] Section 94(1) provides that the acceptance by Parliament of treaties *and other international obligations* (emphasis added) is required when they "contain provisions of a legislative nature, are otherwise significant, or otherwise require approval by the Parliament" under the constitution. No such action was taken at the time when SUCBAS co-operation was established. The issue is included briefly in Government Bill 94/2016, which thus gained a kind of umbrella function of providing legal basis for a wide range of platforms for military and military/civilian co-operation, some of which have been pursued for a considerable amount of time.[65] As it stands today, the main legal basis for the exercise of territorial surveillance in Finland is the Territorial Surveillance

63 A. Jyränki et al., *Konstitutionell Rätt* (2015); A.J.K. Bailes, "Parliaments and National Strategy Documents: A Comparative Case-Study from the Nordic Region," Geneva Centre for the Democratic Control of Armed Forces, *Policy Paper No. 36* (2015).

64 Constitution of the Republic of Finland, 731/1999 (in force 2000); Jyränki et al. (2015) 308–9.

65 Finnish Government Bill RP 94/2016, 15.

Act, which was revised as recently as 2015.[66] Recent amendments of this act strengthen the co-operation between military and civilian authorities responsible for territorial surveillance in Finland as well as place this co-operation under the responsibility of the Ministry of Defence (Section 23). A new section was introduced in 2015 (Section 30a) concerning situational awareness information. This section permits the Finnish armed forces to receive information from the authorities of foreign states or from international organisations when creating such situational awareness information. This would seem to allow to Finnish authorities to receive information, but does not offer a basis for providing foreign authorities with similar information. This was in fact one of the proposals included in Bill 94/2016.[67] These matters affect, and create some confusion for, the distribution of powers between the President (who is also the Superior Military Commander in Finland), the Prime Minister (who heads EU affairs) and the Ministry of Defence and Armed Forces.[68]

Following the Autonomy Act of Åland (Section 58), the Government of Åland shall be informed of negotiations on a treaty or another international obligation if the matter is subject to the competence of Åland, or else when the competence is absent but the matter is of special importance to Åland, the Åland Government "shall be informed of the negotiations, if appropriate". While security and defence matters are not in the competence of Åland and Åland is not a party to any of the underlying international treaties and agreements, matters of demilitarisation have clearly been understood as being of special importance for Åland.[69] While the NEFAB co-operation information was sent by the Finnish Ministry of Transportation and Communications to the Åland Government in spring 2012 alongside a request for comments, no similar procedure was followed with regard to SUCBAS.[70] It is not clear what the reason for such divergence may be.

The SUCBAS Technical Arrangement (Section 8.1) provides that nothing in the arrangement is intended to be in conflict with national or international

66 "Territorial Surveillance Act 755/200 and Amendments in 195/6.3.2015" (2015).

67 New proposed Section 24(d) would allow for international co-operation in matters of territorial surveillance.

68 Jyränki et al., *Konstitutionell Rätt* (2015) 309–22.

69 Poullie (2016); H. Rotkirch, "A Peace Institute on the War-Path: The Application of the Treaty on Open Skies to the Neutralized and Demilitarized Åland Islands and the Powers of the Åland Autonomy," in *Nordic Cosmopolitanism: Essays in International Law for Martti Koskenniemi*, eds. J. Petman and J. Klabbers (2003).

70 Spiliopoulou Åkermark, "The Meaning of Airspace Sovereignty" (2017). The Finnish airspace surveillance has integrated the civil and military sides, something which has not been the case in Sweden, as was confirmed in public debates concerning airplaines flying with turned-off transponders. "Flygledningen vill få tillgång till militär radar," *Helsingborgs Dagblad*, 13 December 2014.

law and agreements. If such a conflict arises, then "national and/or international law and existing agreements and arrangements *between the Participants* will prevail" (emphasis added by authors). This clause does not seem to be fully in line with international obligations concerning demilitarisation of the Åland Islands, which include several countries not participating in the SUCBAS network, among them France, Italy, the Russian Federation as well as the states parties to the Finnish Peace Treaty of 1947 (see Chapter 2). Are states participating in SUCBAS to give priority to agreements and obligations between themselves at the expense of such obligations towards other states? Surveillance co-operation, on land, sea and in the air, has a strong military character, as argued previously. Its implementation can thereby not be pursued within the demilitarised zone of the Åland Islands, even though the possibilities for verifying this for any non-expert in this domain are nearly non-existent. States parties, of course, always have the right to object to practices in contradiction to the legal obligations assumed other by states parties.

Let us imagine a hypothetical situation. A third country, we can call it Country Y, engages in activities along the Finnish border, or in Finnish territory. These activities are observed and shared among the participants. Some of the SUCBAS participants interpret these acts as involving military actors, or as threatening the territorial integrity of Finland. Some of the SUCBAS participants may be more restrictive in their interpretations. This is not at all an unusual situation in international affairs, especially at present, when issues concerning self-defence and the use of armed force are under wide discussion, as has been experienced with regard to the use of force in Kosovo (1999), Iraq (2003) and Syria (2007), to mention but a few examples.[71] How can Finland retain its control over the interpretation of what constitutes a threat to territorial integrity, in order to also uphold the demilitarisation and neutralisation regime of the Åland Islands? How can Finland ensure that its multifaceted international obligations are taken into account by other states parties, authorities and private actors involved in such functional co-operation, and that these norms are respected effectively? How can responsible authorities distinguish between military and civilian actors and activities when both actors and their activities are

71 S. Spiliopoulou Åkermark, "The Puzzle of Collective Self-Defence: Dangerous Fragmentation or a Window of Opportunity? An Analysis With Finland and the Åland Islands as a Case Study," *Journal of Conflict and Security Law* 22, No. 2 (2017). On diverging opinions see e.g. "Which Countries Support and Which Oppose the U.S. Missile Strikes in Syria," *The New York Times* (9 April 2017). Also "A Plea Against the Abusive Invocation of Self-Defence as a Response to Terrorism," *Centre de Droit International, Université Libre de Bruxelles*.

intermingled and often placed under the co-ordination authority of the military, as for instance provided in the Finnish Territorial Surveillance Act (Sections 23 and 24)?

As with co-operation within NEFAB, SUCBAS also entails the sharing of all kinds of situational awareness information between the participating countries. Both initiatives are loosely linked to EU initiatives in the area of territorial surveillance and situational awareness, areas which, however, still fall entirely in the domestic competence of member states. This is also why such initiatives are described as voluntary and optional. While 'transparency' has become the new buzzword, the legal and political questions are accentuated and persistant; they concern the procedures and legal requirements when taking decisions concerning territorial integrity and surveillance, including for exchange of information; the effect of situational awareness co-operation on the assessment of threats to national security; and the effect, legal as well as political, of such co-operation on other international legal obligations, such as those concerning the demilitarisation and neutralisation of the Åland Islands.

What we can observe in such loose forms of co-operation is a shift towards increased priority given to military decision makers and to military responses to security concerns, even when the concerns are not, or, are only in part, of a military nature. We also observe a trend towards a decrease in parliamentary involvement and control on matters pertaining to the very internationalised, network-based and highly technical nature of military-civilian co-operation. These observations confirm, with some of the components, a trend towards 'militarisation', reviewed in the first chapter of this book.

4.4 Finland, Åland and the North Atlantic Treaty Organization (NATO)

Although the question of the demilitarised Åland Islands may not have a major role in the political discretion over NATO membership, decisions made by Finland in terms of security and defence policy also impact the islands. The idea of a possible membership for Finland in NATO divides people both in the Åland Islands and in mainland Finland. Finland has been part of the NATO Partnership for Peace programme since 1994. In 2014, Finland also signed a Memorandum of Understanding on Host Nation Support with NATO. Nevertheless, the current Finnish Government, in power since 2015, has declared that Finland would not apply for NATO membership during their term. With regard to NATO membership, the 2016 Government Report on Foreign and Security Policy states that "[w]hile carefully monitoring the developments in its security environment, Finland maintains the option to seek NATO membership".[72] A 'NATO option' is something that is assumed to exist, although Finland has not been promised such option.

The NATO debate appears constantly ongoing in Finland, and the political institutions further boosted NATO debates in 2016. Firstly, Finland committed to hosting a large NATO-led military exercise (BALTOPS) in summer 2016, which spurred discussion on Finland's relations with NATO and the parliamentary control of NATO exercises. Secondly, the Foreign Ministry commissioned a report on the eventual impact of Finland's NATO membership,[73] which generated much debate about whether it would be possible for Finland to join the Alliance and what Russian reactions might be. It is to be expected, said the report, that the most difficult period would be the application period for joining NATO, when Russian reactions would be harshest in an effort to stop the negotiations. In case Finland was to join the Alliance, it is argued that it should keep the transition period as short as possible in order to minimise antagonism from Russia.[74]

Public support for NATO membership is not very high (approximately 26% of Finns are in favour),[75] but military personnel, in general, have supported NATO membership. In spite of public hesitance, two political parties openly support membership. One of the four largest parties in Finland, the Coalition Party, has traditionally been a supporter of NATO membership, and the Swedish People's Party even declared that Finland should become a NATO member by 2025 in its party programme in spring 2016.

There seems to be widespread political consensus in Finland that Finland and Sweden should join NATO simultaneously, if they choose to do so. In particular, if Sweden were to join NATO, Finland should allegedly follow suit. Swedish membership seems unlikely to occur in the near future since the current (Social Democratic) Swedish Government has announced unequivocally that Sweden would not join NATO, although the debate continues. A Swedish Government-commissioned security report in autumn 2016 considered that Finland should be taken into account while contemplating NATO membership. However, it concluded that Finland should not prevent Sweden from joining.[76] In Sweden, support for NATO member-

72 Prime Minister's Office Finland, "Government Report on Finnish Foreign and Security Policy 6/2016."

73 Finnish Ministry for Foreign Affairs, *The Effects of Finland's Possible NATO Membership—An Assessment*, ed. M. Bergquist et al. (2016).

74 Ibid.

75 H. Vuorela, "Nato-jäsenyyden vastustus väheni—Kanerva yllättyi muutoksen vähäisyydestä," *Maaseudun Tulevaisuus*, 11 January 2017.

76 K. Bringéus, "Säkerhet i ny tid" SOU 2016:57, 150–1.

ship in not particularly strong; about one third of Swedes were in favour of joining the Alliance in summer 2017.[77]

Leading Finnish politicians reiterate the need to have public support for NATO membership. In 2014, Juha Sipilä, who was elected as Finnish Prime Minister the following year, stated that a referendum should be organised before Finland could apply for membership. He has also speculated that a possible membership would decrease Finland's political room for manoeuvre, but would not remove the need to strengthen Finland's own defence capability.[78] In a political speech held at the party conference of the Centre Party in June 2016, he reiterated his view on organising a referendum and emphasised that NATO membership would be an enormous change in Finnish foreign policy.[79] In the speech, he also quoted the NATO report published in spring 2016, stating that "[i]t is, in essence, a question of grand strategy, which has to be considered thoroughly. Small nations do not often change their basic foreign policy guidelines. They are more dependent on continuity than great powers".[80] Without a doubt, NATO membership would be a huge shift in Finnish foreign policy. Traditionally, continuity in foreign policy has been important for Finland, which has sought to maintain its foreign policy principles such as neutrality and non-alignment and only reformulated them when necessary. Neutrality policy was abandoned when Finland joined the EU, and not being a member of a military alliance is held onto, even though Finland committed to the defence clauses of the EU.[81]

Continuity is also crucial in the demilitarisation regime, which is sometimes linked to the NATO debate. Russian reactions to Finnish NATO application or to any attempt at modifying the existing Åland demilitarisation agreements would certainly be critical, to say the least. Tiilikainen speculated in 2002 that in case Finland joined the Alliance, this would not bring about a reformulation of the demilitarisation of the Åland Islands because that would polarise the already tense situation with Russia. Alternatively, Finland could try to "sell everything in one package" and aim at changing

77 J. Pisoni, "Bara var tredje vill att Sverige går med i Nato," *SVT Nyheter*, 3 July 2017.

78 P. Kontio, "Juha Sipilä: Nato-Jäsenyys Rajoittaisi Suomen Liikkumatilaa," *Suomenmaa*, 15 June 2014.

79 J. Sipilä, "*Seinäjoen puoluekokous poliittinen linjapuhe 11 June 2016* [Political Speech at Seinäjoki Party Conference]," 11 June 2016.

80 Finnish Ministry for Foreign Affairs, *The Effects of Finland's Possible NATO Membership: An Assessment* (2016).

81 S. Heinikoski, "Policy Constraints Argued Not to Constrain—Post-Neutrality and Demilitarization in Finnish Foreign Policy Discourses after EU Accession" (Forthcoming) n.d.; S. Heinikoski, "The Åland Islands, Finland and European Security in the 21st Century," *Journal of Autonomy and Security Studies* 1, No. 1 (2017) 8–45.

the demilitarisation regime simultaneously.[82] However, in the 2016 assessment report on Finland's possible NATO membership drafted by Tiilikainen and other experts, the Åland question is not discussed. The only entry on the Åland Islands reads: "[t]he relationship between the international agreements that cover the *sui generis* status of these islands and the undertakings implied in membership need to be examined further".[83] In legal terms, making a reservation concerning the demilitarised islands would probably not be a problem. However, in military strategic terms, it might be challenging, as NATO would have to provide security guarantees for Finland as a whole under Article 5.

The legal implications of a hypothetical NATO membership on the regime pertaining to the Åland Islands have not been officially assessed in Finland. A more general study, with a comparative angle, was commissioned by the Finnish Ministry of Foreign Affairs in 2008.[84] The study concludes briefly that "it seems clear that no change in the special status of Åland would be desirable or necessary due to NATO membership".[85] As often done in similar situations, there is in the study a rather sweeping reference to the Svalbard case.[86] However, it is noted that the special status of the Åland Islands should be taken into account in accession negotiations, since "the principle of flexible movement of troops" could not apply to Åland and there should be "a clear common understanding that the Islands' special status remains unchanged".

Similarly, there was no substantive analysis on the legal implications for Åland of recently adopted Finnish legislation on sending and accepting international assistance.[87] In the first provision of the short new Act 418/2017, general reference is made to the UN Charter and international law. The government bill which preceded the adoption of the act simply noted that Åland is a demilitarised and neutralised region under international law and that this status had been incorporated in the EU legal and

82 T. Tiilikainen, *The Åland Islands, Finland and European Security* (2002).

83 Finnish Ministry for Foreign Affairs (2016).

84 J. Rainne, *Legal Implications of NATO Membership: Focus on Finland and Five Allied States* (2008).

85 Ibid., 48–9.

86 The specific conditions of Svalbard/Spitsbergen and the insufficient knowledge about their legal framework within NATO are shown in T. Koivurova and F. Holiencin, "Demilitarisation and Neutralisation of Svalbard: How Has the Svalbard Regime Been Able to Meet the Changing Security Realities During Almost 100 Years of Existence?," *Polar Record* 53, No. 2 (2017).

87 Finnish Act 418/2017 and Government Bills 72 (2016) concerning provision and acceptance of international assistance and 107 (2016) on assistance in matters falling under the competence of the Ministry of Interior adopted by Parliament in June 2017. See also Spiliopoulou Åkermark, "The Puzzle of Collective Self-Defence" (2017).

political order.[88] As it seems, changes in Finnish policies and legislation are in principle regarded as not affecting the status quo of the Åland Islands.

4.5 The Åland Islands in Regional Security Co-operation as Understood in Finland

In this section, we discuss the demilitarised Åland Islands in the context of Finnish security co-operation, which has gone through a gradual change of emphasis from multilateral rules towards regional security. This process is followed by Finnish efforts for developing extensive bilateral co-operation in the region, including with Sweden, Russia and Estonia. A more thorough analysis of such efforts would be far beyond the scope of the present study. Analysis of the changes in the approach during Finnish EU membership is fruitful for understanding the approach towards the demilitarisation of the Åland Islands, which can be regarded as a part of the larger Finnish foreign policy context. Finland has chosen to stay outside of military alliances but to participate in multilateral arrangements such as the UN. It seems that until 1995, the UN and the OSCE were the main multilateral organisations that mattered for Finnish security, coupled with Finland's independent military capacity. Since 1995, the EU has stepped into the picture, and since 2009, the Nordic Defence Cooperation (NORDEFCO) has also gained in importance in Finnish security policy. Although Finnish defence still relies on national defence capacity, increasing emphasis is put on defence co-operation.

In order to observe developments during Finland's EU membership, the authors analysed Finnish Government reports on security and defence policy as well as related committee documents and parliamentary debates. These white papers are published approximately every four years (since 1995) and prepared mainly by the Prime Minister's Office, the Ministry for Foreign Affairs, Defence Ministry, Ministry for Interior Affairs and the President of the Republic, while consulting a parliamentary monitoring group. Most of the reports have focused on security and defence issues, but the incumbent government decided to divide the reports in two: one on foreign and security policy (2016) and the other on defence (2017). For the purposes of this book, a decision was taken to focus on analysing how the situation has changed during the past twenty years by focusing on three different periods: mid-1990s, mid-2000s and mid-2010s. The first Finnish security and defence policy reports from 1995[89] (focus on security) and

88 Finnish Government Bill 72/2016, 47.

89 Prime Minister's Office Finland, "*Turvallisuus muuttuvassa maailmassa* [Security in a Changing World]," 1995.

1997[90] (focus on defence) as well as related parliamentary debates and parliamentary committee documents form the starting point for the debate.[91] As a comparison, we analyse the security and defence policy report of 2004, the foreign and security policy report from summer 2016 and defence policy report from February 2017, as well as debates and committee documents related to them.[92] It is worth noting that the 1997 and 2017 defence-focused reports build on the estimates of the previous security reports (1995 and 2016, respectively) and are supposed to be consistent with those reports.

This analysis has revealed the sort of role different security concepts and forms of co-operation have had in Finnish security policy since Finland's accession in the EU. Finland is not part of any military alliances, but co-operation in defence matters was presented as of utmost importance in the reports. The European and Nordic communities, which used to promote the idea of a comprehensive security (see sections 4.1 and 4.2), are becoming more competent and engaged in matters of military security.

All reports mentioned the comprehensive security concept as the basis of Finnish policies, but nevertheless mostly focused on military security. The 1995 security policy report was the only one that paid less attention to

90 Prime Minister's Office Finland, "*Euroopan turvallisuuskehitys ja Suomen puolustus* [European Security Development and the Finnish Defence]," 1997.

91 Prime Minister's Office Finland, "*Turvallisuus muuttuvassa maailmassa* [Security in a Changing World]"; Finnish Parliament, "Minutes of the Plenary Session 89/1995 on 1 November 1995," 1995; Finnish Parliament, "Minutes of the Plenary Session on 31 October 1995," 1995; Finnish Parliament, "Minutes of the Plenary Session 31/1995 on 31 June 1995," 1995; Finnish Parliament Committee for Foreign Affairs, "Foreign Affairs Committee Report on the Government Report 1/1995," 1995; Finnish Parliament Defence Committee, "Statement of the Defence Committee on the Government Report 1/1995," 1995.

92 Prime Minister's Office Finland, "Finnish Security and Defence Policy 2004 Government Report 6/2004," 2004; Finnish Parliament, "Minutes of the Plenary Session on 28 September 2004," 2004; Finnish Parliament, "Minutes of the Plenary Session on 29 September 2004 98/2004," 29 September 2004; Finnish Parliament, "Minutes of the Plenary Session on 22 June 2004," 22 June 2004; Finnish Parliament, "Minutes of the Plenary Session on 20 December 2004," 20 December 2004; Finnish Parliament Committee for Foreign Affairs, "Statement of the Committee for Foreign Affairs 4/2004 Government Report on Security and Defence Policy 2004," 2004; Finnish Parliament Defence Committee, "Report of the Defence Committee on the Government Security and Defence Policy Report," 2004; Finnish Parliament, "Minutes of the Plenary Session on 21 June 2016," 21 June 2016; Finnish Parliament Committee for Foreign Affairs, "Government Report on Finnish Foreign and Security Policy 9/2016," 22 November 2016; Finnish Parliament Administration Committee, "Statement on the Government Report on Foreign and Security Policy HaVL 40/2016," 2016; Finnish Parliament, "Minutes of the Plenary Session on 21 December 2016," 2016; Prime Minister's Office Finland, "Government Report on Finnish Foreign and Security Policy 6/2016"; Prime Minister's Office Finland, "Government's Defence Report VNS 3/2017"; Finnish Parliament, "Minutes of the Plenary Session on 8 March 2017."

military security and more attention to stability and conflict management. From the government's perspective, it may be effective to employ security threats to justify more funding for defence in line with the securitisation perspective (see the Introduction to this book).

Multilateralism is seen positively in the reports and is something which is important to promote, but it is occasionally connected to a sense of obligation and restriction for Finnish defence. Examples of this include statements that no military exercises could be held in the demilitarised Åland Islands as well as vivid discussions on the ban on anti-personnel mines. In both cases, however, compliance with international agreements was given priority before defence considerations. It seems thus that despite increased emphasis on regional security, the need to comply with international law has not decreased in importance. The role of the different arrangements from the 1990s to the present is visible in the following table.

As can be seen in the table below, the role of these arrangements in Finnish and global security has changed over the years. The development is briefly summarised here. Although belief in the effectiveness of the

Table 4.1 Approach to multilateral and regional arrangements from the mid-1990s until today in Finnish security and defence policy government reports.

Arrangement	*1995/1997*	*2004*	*2016/2017*
Demilitarisation of the Åland Islands	Established status but considered a challenge for defence	Considered to increase security and confidence in the region	Established status, does not hinder defence co-operation
European Union	Comprehensive security organisation that should not become a defence union	The most important actor in responding to cross-border threats	Should become a stronger security community and co-operate with NATO
UN	Multilateral organisation dealing with international rules and crisis management	Multilateral co-operation key to prevent and solve global problems and regional conflicts	Important multilateral arrangement with regard to comprehensive security, but Security Council incapable
Nordic Defence Cooperation	Non-existent but worth considering	Co-operation under NORDCAPS in crisis management	NORDEFCO important, also bilateral co-operation with Sweden emphasised

multilateral UN was reduced in the latest foreign and security policy report, the demilitarisation of the Åland Islands was not questioned. It was considered a challenge for defence by the Defence Minister, but the government stated that no changes in the status were planned.[93] Regional arrangements seemed to constitute the most crucial factors in Finnish defence throughout the analysed period, and co-operation was promoted in the EU, in the Nordic context and bilaterally. The entire idea of independent defence has thus changed considerably; Finland did not have to assume sole responsibility, but security in the Northern and European contexts were seen to be inextricably interlinked. Cyber security seemed to be one of the most important fields of co-operation in the latest reports. In the 2004 report, cyber security was considered a criminal matter to be handled by the police and the criminal justice system. The 2017 Defence Report, in turn, discussed the cyber field as a defence branch, in addition to air, land and sea defence. This could be considered securitisation and militarisation of cyber threats, transferring new non-military issues to the state military security agenda. Such threats were also considered to be more effectively addressed with regional defence co-operation rather than multilateral rules and norms or police co-operation.

On the basis of the reports, it can be concluded that the role of the EU as the most important security-enhancing mechanism in the Finnish nearby area has remained constant, but more far-reaching defence measures are supported in the latest reports. Nordic and European co-operation are seen as the most important fora, but the co-operation had enlarged from comprehensive security issues also, explicitly, to military ones. Perhaps the most drastic change relates to Nordic defence co-operation. In 1995, the Nordic countries were mainly described as a value community, where co-operation was conducted in non-security issues. In 2004, Nordic co-ordination in crisis management emphasised the changed security concept and focused on crisis management rather than territorial defence. The 2016 report emphasised NORDEFCO and bilateral co-operation with Sweden as the closest defence co-operation partners of Finland. As regards the EU, no common defence co-operation arrangements were in sight in the 1990s and politicians often wanted to avoid them taking place.

The demilitarisation of the Åland Islands can be seen in the light of this change in emphasis: the established status of the multilateral arrangement is acknowledged, but it became important to highlight regularly that the status does not hinder Finnish defence co-operation. Although not all

93 Finnish Parliament, "Minutes of the Plenary Session on 20 Oct 2016: Oral Question on the Demilitarisation of the Åland Islands."

reports mentioned demilitarisation, the Åland Islands always received some attention in the Finnish parliamentary debates. In these debates, one can observe how the approach towards demilitarisation differs depending on whether the speaker focuses on international law or defence aspects. A good example of the different perspectives was a Finnish parliamentary debate on 20 October 2016, when a Left Alliance MP presented an oral question to the Defence Minister concerning the latter's expressed apprehension caused by the demilitarisation of the islands. The Defence Minister and the Foreign Minister (both representing the nationalist Finns Party at the time) said they concurred with the government's stance, which upheld the status quo. However, whereas the Defence Minister introduced the challenges of defending the islands, the Foreign Minister stated that "[t]he Finnish stance is the same it has been for a long time, and I think it is a good one. We respect international agreements, we hold onto what has been agreed. This is a point of honour for Finland".[94] Whereas the Defence Minister considered the agreements a burden for defence preparation, the Foreign Minister regarded them as a question of honour; this division between defence and normative concerns is visible also in other debates on the status, as the authors are about to illustrate.

In 1995, the parliamentary debates discussed the Government Report on Security Policy also in the context of the demilitarisation of the Åland Islands, but demilitarisation was not discussed in the actual report. In the debates, the Swedish People's Party mentioned that the demilitarisation of the Åland Islands was not mentioned in the report as they considered should have been the case. The Åland example was also promoted as a model for European crisis management by the Chair of the Swedish People's Party.[95] The Swedish People's Party also wanted to emphasise that neither the government nor the parliamentary committees had considered the Åland Islands a problem from a security perspective; the underlying assumption therefore was that demilitarising an area could be a security problem. The only Ålandic MP in the Finnish Parliament also addressed the question of demilitarisation, arguing that the lack of attention meant that there were no changes in the status of the Åland Islands.[96]

The Baltic Sea region was often referred to in all reports. The role of the CBSS and of the security in the Baltic Sea was explicitly discussed in 1997; one reason for the central significance of the Baltic Sea for Finland was "the responsibility for the defence of the strategically important Åland Islands,

94 Ibid.
95 Finnish Parliament, "Minutes of the Plenary Session 31/1995 on 31 June 1995."
96 Ibid.

demilitarised during peace time".[97] The Åland Islands were thus viewed from the perspective of defence and with reference to peace time demilitarisation. Finland holds rights and obligations concerning the neutralisation of the islands in a war, for example, by preventing attacks with sea mines and by not using the islands for any war purposes. Sea mines were also mentioned in the 1997 report, stating that the "defence of the Åland Islands is based, to a large extent, on the mining ability of the marine forces, since no defence preparations may take place in the demilitarised zone".[98] Demilitarisation was therefore accepted as an obligation binding Finland, and the defence of the islands had to be considered within the limits of the agreements.

Perhaps due to the focus on defence of the 1997 report, the preliminary debate on the report also emphasised further the defence aspects related to demilitarisation. The demilitarisation of the Åland Islands was brought up in the Swedish People's Party's group speech; they stated that the status had to be taken into account in defence policy. In the same debate, another MP also discussed the problems of defending the demilitarised Åland Islands.[99] One MP did refer to the treaties but suggested abolishing demilitarisation because he deemed that there were not many signatory states left.[100] This illustrates fittingly the lack of knowledge on the agreements among Finnish politicians. All the signatory states do exist and are bound by the agreements; the governments and borders of states may have changed since the 1921 Convention, but that does not alter their legal obligations.[101]

The 2004 report did not mention the Åland Islands, but the Ålandic MP requested the committees to take a stand on the status in their discussions.[102] Indeed, the Foreign Affairs Committee did mention demilitarisation (and neutralisation) in their statement, stating that "Åland is a demilitarized and neutral [sic] area, which the Committee considers an arrangement contributing to increasing security and confidence in our region".[103] The Defence Committee also mentioned the two demilitarisation agreements (of 1921 and 1940) and reminded others that both agreements bind all parties,

97 Prime Minister's Office Finland, "*Euroopan turvallisuuskehitys ja Suomen puolustus* [European Security Development and the Finnish Defence]" 1997.

98 Ibid.

99 Finnish Parliament, "Minutes of the Plenary Session 28/1997 on 17 March 1997," 1997, 685.

100 Finnish Parliament, "Minutes of the Plenary Session 29/1997 on 18 March 1997," 1997.

101 Hannikainen (1994).

102 Finnish Parliament, "Minutes of the Plenary Session on 28 September 2004."

103 Finnish Parliament Committee for Foreign Affairs, "Statement of the Committee for Foreign Affairs 4/2004 Government Report on Security and Defence Policy 2004."

including Finland.[104] It was rarely acknowledged in the parliamentary committees or debates that there are indeed more agreements besides the two most well-known ones.

After 2004, all the reports included a reference to the demilitarised islands, perhaps in order not to have to explain its absence. The 2016 report stated that "[t]he Province of Åland Islands has a recognised status under international law. This does not prevent Finland from intensifying defence cooperation within the European Union, with international organisations and in the Nordic context".[105] A similar entry had been included in the reports of 2009 and 2012, but this time it was complemented with the Nordic context. The demilitarisation of the Åland Islands was considered to have a stable role based on international law, and the parliamentary debates related mainly to the issue that the Åland Islands did not create any problem for Finnish defence or defence co-operation. One issue to be noted is that the Defence Minister had raised his concerns related to the defence of the Åland Islands in autumn 2016 in the media. This was not recorded in the debates, but due to the vivid media debate on the status, the parliamentary Foreign Affairs Committee specified in its report that other states had not questioned the status and it was in the Finnish interest to guarantee the demilitarised status, as well as also having responsibility for its defence.[106]

The Åland Islands did not appear in the 2017 report, but the statement of the Foreign Affairs Committee of the parliament considered it worth mentioning with a very similar formulation as that of the previous report. The Defence Committee, in turn, did not discuss demilitarisation in its report. Two weeks after the publication of the Defence Report, the Defence Minister wrote a short article stating that "[d]ue to being demilitarised, the Åland Islands, however, create a challenge for Finland in defence terms".[107] He has made similar statements from time to time during his entire term in the post, but he has, together with the foreign policy leadership, assured the public that the status is stable. Bringing up defence challenges related to the islands may reflect a willingness of the Defence Minister to have a say in issues that normally belong to the competence of the Foreign Ministry. Another implication of the desire to have more defence emphasis in security matters is the fact that the incumbent government decided to publish separate reports,

104 Finnish Parliament Defence Committee, "Report of the Defence Committee on the Government Security and Defence Policy Report," 2004.

105 Prime Minister's Office Finland, "Finnish Foreign and Security Policy 2016 Government Report 6/2016."

106 Prime Minister's Office Finland, "Government Report on Finnish Foreign and Security Policy 9/2016."

107 J. Niinistö, "Itämeren geostrateginen merkitys kasvussa," *Pulloposti*, No. 9 (2017).

one on foreign and security policy and one on defence. However, the fact that the 2017 Defence Report did not mention the Åland Islands implies that demilitarisation was acknowledged rather as a multilateral agreement, that is, as a foreign policy issue, only discussed in the 2016 Foreign and Security Policy Report. In the media debates concerning the matter, the focus still seems to be on regional security implications.

4.6 Demilitarisation Between Regionalism and Multilateralism

In the present chapter, the authors have looked at security co-operation and its relation to the Åland Islands. Some consistency in the Finnish approach can be observed at least since the end of the Cold War: security is not achieved through isolation but through co-operation, whether it is multilateral or regional. Demilitarisation serves as a prime example of a functioning multilateral arrangement, whereas Nordic and European regional co-operation contributes to building regional security. In this last section of the chapter, the aim is to draw some parallels between regionalism and multilateralism. They follow, to some extent, a similar logic aimed at preventing conflicts and war. However, where (successful) regionalism usually requires some level of similarity, international law may also be successful in case of conflictual parties. The demilitarisation regime, established in 1856, was in the mutual interests of all parties, representing different sides in the Crimean War: Great Britain, France and Russia. Although the demilitarisation was 'imposed' upon Russia, it also preferred to focus resources in the Black Sea. Looking at demilitarisation from a contractual point of view, we see that it involves both realism and idealism. While it is in the interests of all parties to maintain their sovereignty and prevent conflicts, normative ideals of maintaining peace through agreements are also at play.

The emphasis on the different aspects of demilitarisation fluctuates with time. The first demilitarisation agreements were certainly informed by realist concerns after a recent war. In 1921, however, the demilitarisation was linked to collective security, to the autonomy arrangement and to minority rights in addition to military viewpoints, thus giving it a more idealistic character. It was not warring superpowers that decided on demilitarisation after the First World War, but an international multilateral organisation, the League of Nations, drafting the agreements from a wider perspective. The idea that peace could and should be dealt with through internationalised co-operation, and that war should be limited and controlled, had created another idealist string of logic. The 1940 agreement between Finland and Russia, on the other hand, seemed to reflect a more realist and military perspective, and so did the open-ended mention in the Paris Peace Treaty after the Second World War. Justifying something on the basis that it has been stipulated in agreements does not make it unquestionable, but an argument pointing at the long-term stability-enhancing effect of international agreements is a heavy empirical and functional statement.

The demilitarisation of the Åland Islands is closely connected to what is happening in the region around it and worldwide. In times of peace, emphasis is on international law, whereas fear and securitisation can lead to thinking more in security terms, as we have discussed. The crucial issue with regard to the demilitarised and neutralised Åland Islands and other regional and multilateral arrangements seems to be the quest for stability through mutual formalised legal commitments. Indeed, the very idea is that such commitments are even more necessary and meaningful on controversial or disputed matters.

5 Outlook and Conclusions

In this book, the concepts of demilitarisation, security and militarisation have been examined. Furthermore, the authors have looked at the legal frameworks of the contingent solutions pertaining to the Åland Islands and have concluded that the rules are considered still valid and relevant even though many of the circumstances surrounding them have been challenged and reconfirmed over time. Two fields of such evolution have been examined in more detail in Chapters 3 and 4, namely the law of the sea and navigational rights and then regional security co-operation, respectively. In this last chapter, the overall theoretical and principled questions of an assessment of the demilitarisation and neutralisation of the Åland Islands are discussed.

5.1 Peace Through Law?

The idea of collective security as epitomised today in the UN Charter rests on the wish to prevent war and restrict the use of force to a minimum. This idea was quite radical as it meant that security considerations had a collective and international dimension and did not only rest upon the premises of national interests and national security. As we saw in section 2.2, the League of Nations Secretariat, through Anzilotti and Kaeckenbeeck, had drawn the attention of concerned states to the relationship between the demilitarisation regime of a particular territory, on the one hand, and collective security, on the other.

However, the idea of collective security did not replace or erase the institution and rules of self-defence. The UN Charter is a condensed regulatory summary of international relations in the light of two world wars and the long-lasting evolution of a regulatory framework for international affairs which we call international law. The UN Charter aims at maintaining "peace and security" and enhancing international co-operation, based upon the premises of sovereign equality of states coupled with the obligation of these sovereign states to solve disputes by peaceful means and refrain from the threat or use of force. This system was more thorough, but not radically new, as compared to the one that existed in the League of Nations era. Matters of war and peace became part of the internationalised and institutionalised competence of the League of Nations in 1919 as a consequence of the devastation and trauma brought by the First World War. Article 10 of the Covenant of the League of Nations provided:

> The Members of the League undertake to respect and preserve as against external aggression the territorial integrity and existing political independence of all Members of the League. In case of any such aggression or in case of any threat or danger of such aggression the Council shall advise upon the means by which this obligation shall be fulfilled.

In fact, the League of Nations was perhaps more of a system of collective self-defence than prohibition of the use of force, and it was more reminiscent of a military alliance or military mutual assistance clause, albeit offering also, in tandem, a new and structured system of peaceful dispute resolution through the Permanent Court of International Justice or through arbitration. A prohibition of wars of aggression was incorporated later in the so-called Briand-Kellogg Treaty in 1928, this time with the United States among the signatories, but outside the framework of the League of Nations and without any system of legally binding decision-making in collective security matters.[1]

In this process of trying to regulate the use of force and prevent war, the internationalised tool of demilitarisation and neutralisation was an early component. This institution 'simply' limits the presence and use of military force in a geographically defined area through agreement or through unilateral commitment. This form of limited demilitarisation is distinguishable from the overall demilitarisation imposed upon entire political and legal orders, such as was the case in Japan or Germany after the Second World War.

It is evident that law cannot in itself eradicate war. This would be an unrealistic expectation. Even more so, law cannot replace politics. Law, whether it be national or international, can, at best, be seen as a self-limitation and a normative pull in the conduct of politics, following ideas originally put forward in the rule of law debates of the 19th century. More than being a

1 General Treaty for Renunciation of War as an Instrument of National Policy, *League of Nations, Treaty Series*, 1929, Vol. XCIV, 57.

self-limitation, law is the outcome of politics and, in the democratic state, as well as in international institutions, is expected as such to follow conditions of legitimacy and legality. Finally, law is a possible tool for the conduct of politics, of which Finland is an important example, both in using the legal avenues under the Russian empire era, as well as in placing considerable emphasis on constitutional legality and international law during its one hundred years as an independent republic (1917–2017). The emphasis on law and on legality is, in this case at least, a way to overcome societal polarisation at the domestic level (as became evident in what was effectively a civil war of 1918), as well as pressures in an internationally polarised situation, such as the conditions during the Cold War when Finland had to balance between the 'East' and the 'West'. The above arguments underline the importance of democratic control in matters pertaining to security and the use of force.

Another attribute of the tool of demilitarisation is that it both limits as well as reinforces the idea of sovereignty over the territorial zone concerned. This can partly explain the somewhat heterogenous views expressed about demilitarisation by various actors. Demilitarisation as a tool of international affairs allows for regular reminders, of different kinds, of the authority exercised by the state, in this case Finland, and thus of the sovereignty of Finland over the islands. This may take the form of communication by Finland to other states parties to the relevant international treaties (as we saw happened during the world wars but also more recently when the cartographic co-ordinates were updated; see section 2.5), notifications to the depositaries of such instruments (as was the case of Estonia in 1992 under the 1921 Convention and the Russian Federation in relation to the bilateral treaty of 1940)[2] or reference to the status of the Åland Islands in international negotiations (as was done in the process of ratification of the Lisbon Treaty by Finland after acceptance also by the Åland Parliament; see Chapter 4).

The regime of demilitarisation and neutralisation does not present an absolute protective shield for the Åland Islands in case of war. As previously discussed, there were some fortifications built during the wars which were later destroyed; there were bombings, albeit of a limited scope; and there was great apprehension also on the Åland Islands both at the time of the Russian revolution(s) and the Finnish war of independence, which was equally a civil war, as well as during the various stages of the Second World War.

The Åland Islands regime has contributed to providing a strong international identity for Finland and been one of the reasons Finland and Sweden

2 See Chapter 2 of this present book. Both these confirmations took place within less than three weeks in July 1992. The Protocol was signed on the 11th while the Estonian declaration was registered on 27 July 1992.

have insisted on respect for international law.[3] This regime relies on the idea that a state of war is distinguishable from a state of non-war, and that it is, furthermore, possible and meaningful to distinguish between military and civilian actors, organisations, ships, airplanes, drones, persons and so on. The distinction between war and peace, civilian and military, are preconditions for basic ideas of international law, such as the protection of civilians during armed conflict and the identification of legitimate military targets. The same distinction is reflected institutionally in the different tasks of the police and the army and the separate ministries responsible for them.

In Chapter 1 the concepts of militarism and militarisation were discussed. As argued throughout this study, there is today an increased fusion and confusion between military and civil rules, actors and activities. Co-operation such as the Proliferation Security Activity (PSI, section 3.3.2), the Sea Surveillance Co-operation in the Baltic Sea (SUCBAS, section 4.3), recent Finnish legislation on international assistance (Act 418/2017, discussed in sections 4.1 and 4.4) and discussions on cyber war are examples of such fusion. The increased blurring of the distinction between peace and war renders, in turn, the distinction between what is permitted and what is prohibited under the demilitarisation regime, as well as what constitutes innocent passage, which is even more problematic. While similar issues may always have occurred, the speed and profound impact of modern technology on our lives has not been matched by an equally speedy and profound adjustment and adaptation of the various domestic and international legal frameworks. One concrete outcome is the tendency, both in law and in policy, of using military means when responding to non-military threats, including at the domestic level, for instance in the case of natural disasters or major accidents at sea in Finland.[4] The preamble of the 1921 Convention on the Non-Fortification of the Åland Islands aims at ensuring that the islands "never become a cause of danger from the military point of view". From 1856 to 1921, further to the 1940 bilateral treaty between Finland and the Soviet Union, the regime of the Åland Islands was reconfirmed and widened to cover not only establishments but also military activities and presence in peace time and at war time, with a certain difference in the regulations applying for Finland and for other states. This regime therefore also covers the 'grey zone' we experience today between war and peace. It is up to all states concerned, particularly Finland, as sovereign state, to ensure that the

3 M. Koskenniemi, "Diplomats, Professors, and Then Some: Notes for a History of International Law in 20th Century Finland," *Nordic Journal of International Law* 85, No. 4 (2016) 322–33.

4 "Final Report of the Committee for a Revision of the Self-Government Act for Åland," Ministry of Justice, Report 33(2017) "*Ålands självstyrelse i utveckling*", 38.

islands are not used for any military purposes and are not perceived as a cause of danger in the region.[5]

5.2 The Concept of Sovereignty Reshaped

The creation of nation-states, alongside the creation of and struggles between empires and processes of colonisation, stretched over many centuries and created new forms of polycentric interactions of societies beginning in antiquity and the Middle Ages. Territory was often deemed to be the property of the monarchy or the dynasty. Kings and their government apparatuses established spatial limits and new understandings of territorial control not only in Europe but around the world.[6] With inspiration in the work of 14th century Maghrebi historian Ibn Khaldun, it has been argued that, at times of peace, the structure of dynastic space followed a concentric scheme, reminiscent of the ripples created on the surface of water by a falling stone. Dynastic power blurred in the most distant peripheries, creating 'variable geometries', or differentiation in terms of power relations, functional focal areas and institutional and procedural solutions.[7] This situation was perhaps not so different from present-day understandings of 'multilevel governance'. However, property gathered should, following Ibn Khaldoun once more, be "invested in the welfare of the subjects and used to give them what is due to them and prevent them from need".[8] Such variable geometry was relevant also in the more scarcely populated peripheries of the various empires, such as the northern part of the Baltic Sea as well as large parts of North Africa, beyond the magnet cities, such as Alexandria and Tunis, along the Mediterranean, or Uppsala and Åbo on the northern side of the Baltic Sea. It was also important for the Åland Islands, the archipelago scattered in the middle of the Baltic Sea, where shores are "close enough to permit easy contact, but far enough apart to allow societies to develop distinctively under the influence of their hinterland as well as of one another".[9]

5 Preamble of the 1921 Åland Convention.

6 F. Ben Slimane, "Between Empire and Nation-State: The Problem of Borders in the Maghreb," in *Mediteranean Frontiers: Borders, Conflict and Memory in a Transnational Word*, eds. K. Nicolaidis and D. Bechev (2009) 35–55.

7 Ibid., 41. On early conceptions of territory and border, see further S. Elden, *The Birth of Territory* (2013).

8 Cited from P. Leuprecht, *Reason, Justice and Dignity: A Journey to Some Unexplored Sources of Human Rights* (2011) 61.

9 D. Abulafia, "Mediterranean History as Global History," *History and Theory* 50, No. 2 (2011): 220–28, at 28 but describing the Mediterranean conditions.

Sovereignty and territorial sovereignty are core concepts still today and perhaps even 'clichés' of international lawyers.[10] Sovereignty is increasingly understood as a multi-layered notion that can easily be both misunderstood as well as misused, as evidenced by the debates of the Brexit-referendum campaigns, but also more generally in European integration.[11] The link between sovereignty, statehood and territory is close; none of those terms has, however, an absolute and closed content.[12] The long intellectual engagement with the concept and functions of sovereignty dates back at least to Jean Bodin's *Les six livres de la république* (1576), through Thomas Hobbes' *Leviathan* (1651), all the way to contemporary analyses. Grimm argues: "just as sovereignty cannot claim a timeless meaning, it is not a concept that remains the same irrespective of location. Instead, it can be assumed that its content also changes from country to country".[13]

Ensuring the security of the population has been an important task and function of the sovereign. Hobbes thought of sovereignty as indivisible, with the goal of efficient political rule for the protection of the people. Sovereignty, following Hobbes, had six core elements which were all invested in 'the sovereign': legislation, adjudication, the making of war and peace, allocating offices, reward and punishment and assigning ranks and honours. National and international aspects of security are today understood as equally important, while issues and decisions on war and peace retain their core position in domestic as well as in international political decision-making. The democratic struggles of the 18th and 19th centuries added the claim of 'popular' to the concept of sovereignty, leading incrementally to the idea of the democratic constitutional state with all its connotations about the separation of powers and the principles of constitutionally entrenched legality and parliamentarism. This long conceptual and social journey means that we, today, require democratic decision-making based on the principle of legality with regard to the use of force, including decisions on war and peace. This also entails that decisions concerning demilitarisation and neutralisation are to be taken following the constitutional and legal frameworks applying at the national and international levels.

10 K. Knop and S. Marks, "The War Against Cliché: Dispatches From the International Legal Front," in *Sovereignty, Statehood and State Responsibility: Essays in Honour of James Crawford*, eds. C. Chinkin and F. Baetens (2015) 3–22.

11 S. Spiliopoulou Åkermark, "The Puzzle of Collective Self-Defence: Dangerous Fragmentation or a Window of Opportunity? An Analysis With Finland and the Åland Islands as a Case Study," *Journal of Conflict and Security Law* 22, No. 2 (2017).

12 D. Grimm, *Sovereignty: The Origin and Future of a Political and Legal Concept* (2015).

13 Ibid., 4.

New understandings of what constitutes a threat have deepened co-operation in the field of collective security, without affecting, however, the idea of self-defence as endorsed in customary law and in Article 51 of the UN Charter. Currently, the development of European co-operation in this field poses questions as to the decision-making and legitimating processes on the use of armed force in participating states, as well as about the relationship between the collective security order under the UN Charter and other particular international legal rules, such as the demilitarisation and neutralisation of the Åland Islands. The interpretation and implementation of the 'mutual assistance clause' (Article 42.7 of the Treaty on the European Union) and the 'solidarity clause' (Article 222 of the Treaty on the Functioning of the European Union) is still open and controversial, as evidenced by the diverging directions taken after the terrorist acts in Paris in November 2015 and shortly thereafter the bombings in Brussels in March 2016. Finland has been an eager supporter of such EU security co-operation, even more so perhaps after the insecurities on NATO priorities and the uncertain directions of the US administration.[14]

Such international normative and political developments have also had considerable effects on the Finnish legal system. One of the most recent and profound developments of this kind is the introduction before parliament of draft legislation concerning international assistance involving the use of force to and from Finland, discussed in section 4.4. Following the new Act 418/2017, Finland could not only be involved in more types of military action abroad, but also accept foreign troops on Finnish soil, deviating for the first time from the long-lasting practice that followed the end of the Second World War and the adoption of the UN Charter.

There are risks attached to such trends. Accusations and perceptions of an alleged democratic deficit directed against international organisations, including the EU and the UN, and more broadly directed against international law, can be further accentuated under circumstances of increased interventionism, military presence and activities by foreign actors around the globe. Even if such accusations tend to downplay the realities of democratic challenges and deficits also at the nation-state level, our attention is diverted from the need to theorise and institutionalise legitimating grounds, processes and institutions that are adaptable and appropriate for a globalised and interconnected world.[15]

14 Spiliopoulou Åkermark, "The Puzzle of Collective Self-Defence", (2017).

15 M. Goodhart, "Europe's Democratic Deficits Through the Looking Glass: The European Union as a Challenge for Democracy," *Perspectives on Politics* 5, No. 3 (2007) 567–84; D. Innerarity, *The Transformation of Politics: Governing in the Age of Complex Societies* (2010).

The Åland solution of the early 20th century, with its elements of identity protection, autonomous political institutions and demilitarisation, contributed to the entrenchment of the ideas of legality, constitutionality and international co-operation not only in Finland, between Finland and Sweden and in the Baltic Sea region, but also globally. While some aspects of that complex solution may today seem old-fashioned or even outdated, they form all in all the outcome of a 'bargain', encapsulated in legal agreements and documents which addressed the political, geopolitical, cultural, security and even financial aspects of this complex and multifaceted situation. This solution involved a measure of satisfaction as well as disappointment for all actors involved. They were all committed, however, to seeing it through and respecting the outcomes of domestic and international settlements. Above all, this solution, including the demilitarisation and neutralisation of the islands, allowed both Finland and the Åland Islands to pursue their lives moving away from heteronomy to the extent possible. Grimm's argument of a move away from heteronomy alludes to aspirations of self-determination and autonomy, in the literal meaning of being governed by one's own decisions, and has the advantage of focusing on the movement rather than the achievement of a status.[16] The hegemonic 'other' can vary over time. It can be another country, interfering or occupying, a military force, an international institution or a private actor, national—such as a political party or a dictator—or international—such as an international corporation. Any such actor can capture the democratic decision-making, the collective self-formation, which allows people to believe that they have some measure of control over their lives and can take the essential political decisions directing their future.[17]

In the case of the Åland Islands, Finnish sovereignty has been modified at two levels: firstly with reference to the demilitarisation and neutralisation of the islands, and secondly, Finnish sovereignty has been modified through the recognition of an autonomous status for the archipelago. Historically, as well as legally, the institutions of autonomy and of demilitarisation and neutralisation of the Åland Islands have evolved and have been legally entrenched at different points of time and through separate legal instruments. The demilitarisation has its origins in the 1856 agreement between France, Great Britain and Russia. Neither Sweden, nor Finland, which was at that time part of the Russian Empire, were parties to that agreement (see Chapter 2). The complexity of the situation increased in 1921, after the proclamation of the Republic of Finland, the Russian revolution(s) and the

16 Grimm, *Sovereignty: The Origin and Future of a Political and Legal Concept* (2015).
17 M. Koskenniemi, "What Use for Sovereignty Today?," *Asian Journal of International Law* 1, No. 1 (2011) 61–70.

dispute between Finland and Sweden which was brought before the League of Nations. As we have seen in Chapter 2, both commissions appointed by the League of Nations saw the reconfirmation and broadening of the demilitarisation regime as a necessity. The decision of the LoN Council of 24 June 1921 referred both to the additional guarantees needed in the autonomy law, which had been introduced already earlier by the Finnish Parliament, as well as to the negotiation and adoption of an international agreement concerning the "non-fortification and neutralisation" of the archipelago. While the council decision of June 1921 envisaged that the 1856 treaty would be "replaced by a broader agreement", eventually the 1921 Convention came to complement rather than replace the previous agreement, as explicitly explained in its preamble.

While the self-government authorities of the Åland Islands do not have any direct and legal competence in military matters, there are at least three ways in which the sphere of self-government and the sphere of demilitarisation and neutralisation meet. *First*, there is the perspective of the integrity and coherence of international legal obligations. From the point of view of international law and the 1921 settlement, Finland and all other parties need to respect all parts of their undertakings. This is particularly important for small countries like Finland, who cannot rely primarily and fully on might but need rather to proceed on the grounds of right and diplomacy.[18] Since the end of the Second World War, Finland has tried to strike a balance between an adequate and self-sufficient defence and an active and rules-based diplomacy within a context of neutrality or non-alliance. This balancing act is currently under strain and is undergoing reassessment in the broader context of deepened European security co-operation, the future directions of collective security, terrorism, cyber security and the increased unpredictability of relations between the great powers and between the EU and other parts of the world. Adherence to international law, including to some of the very early, even foundational treaties ratified by Finland in the early days of its life as an independent republic, becomes thus an important element in this tradition.

Secondly, the exemption of the Ålanders from military service is formally and legally not part of the demilitarisation regime, even though it may underscore and may be contributing to the guarantee that no military

18 See e.g. B. Thorhallsson and A.J. Bailes, "Do Small States Need 'Alliance Shelter'? Scotland and the Nordic Nations," *Security in a Small Nation—Scotland, Democracy, Politics*, ed. A.W. Neal (2017) 52; B. Thorhallsson and A. Wivel, "Small States in the European Union: What Do We Know and What Would We Like to Know?," *Cambridge Review of International Affairs* 19, No. 4 (2006) 655.

force is situated on the islands. In the currently valid 1991 Autonomy Act of Åland it is provided:

> A person with the right of domicile may in place of conscription for military service serve in a corresponding manner in the pilotage or lighthouse services or in other civilian administration.
>
> Service in the pilotage and lighthouse services shall be as provided by a State Act after the Åland Parliament has been reserved an opportunity to submit an opinion on the matter. Service in other civilian administration shall be provided by a State Act with the consent of the Åland Parliament. Until such service has been organised, the residents of Åland referred to in paragraph 1 shall be exempt from conscription for military service.
>
> Paragraph 1 shall not apply to a person who has taken up residence in Åland after having reached the age of twelve years.

Since no specific legislation has been enacted providing for alternative civilian service, residents of Åland who fulfil the above criteria are in principle exempted from conscription.[19] A similar provision had been introduced in 1920 and in the first autonomy act. The very first legal commentary on the autonomy legislation by Tollet and Uggla, who had been instrumental in its development and adoption, noted that there was wide public opinion at the time which did seek to justify a full exemption from all such service for Ålanders on the basis of the demilitarised and neutralised status of the islands.[20] Some years later, during the discussions between Finland and Sweden on the so-called Stockholm plan (1938–1939), the two issues were again brought together. This junction culminated in the Farmers' March that took place on Åland in October 1938 (see section 2.4).

Finally, the autonomous authorities' commitment to the Åland demilitarisation and neutralisation has been upheld throughout the end of the Cold War, negotiations for membership to the EU in the 1990s and various controversies such as the ratification of the Open Skies Treaty in 2002 (cf. Chapter 2).[21] Institutionally, the self-government committee of the Åland Parliament (*Självstyrelsepolitiska nämnden*) has consistently addressed issues of demilitarisation in its workings. In October 2015, the Åland Government approved, following parliamentary discussion in the Åland Parliament,

19 See section 2.4 of this present book.

20 A. Tollet and J. Uggla, *Lagstiftningen angående självstyrelse för Åland: Jämte tillhörandeförfattningar* (1930) 119.

21 Y. Poullie, *Åland's Demilitarisation and Neutralisation at the End of the Cold War* (2016).

including its self-government committee, a *Policy for the Demilitarisation and Neutralisation of Åland* with the subtitle "A handbook for the officials of the Åland government".[22] Such a document cannot modify the allocation of legal competences under the Autonomy Act and the Constitution of Finland. In addition to increasing knowledge and awareness on the demilitarisation and neutralisation regime on the islands, it would seem that this document is a tangible reminder of the wish of the autonomous authorities (Åland Government and Parliament) to be considered as legitimate interest holders in all such matters, primarily vis à vis the Finnish Government and Parliament, but also in relation to all other countries and organisations with rights and obligations under the demilitarisation and neutralisation regime.

The Finnish Ministry for Foreign Affairs established a Contact Group in 1998 between the ministry and the Åland Government. Its institutional hub is at the legal department of the Ministry for Foreign Affairs and is headed by the director general of its legal service. While issues of demilitarisation are only one of the areas of the group's mandate and work, in addition to many others, such as treaty issues and EU affairs, the Contact Group has initiated or facilitated several discussions, seminars and reports over the years pertaining to the demilitarisation and neutralisation of the islands.[23]

All this illustrates how domestic and constitutional aspects, institutions and legislation are intermingled with international obligations and institutions of a bilateral and multilateral nature, not least within the EU. The regime of demilitarisation and neutralisation of the Åland Islands limits and expands Finnish sovereignty at the same time. This is one of the paradoxes of the demilitarisation and neutralisation solution. It limits Finnish sovereignty with reference to restrictions on military presence in the archipelago as well as the Åland Parliament's right to be heard in matters of special importance to the islands under the Autonomy Act. It confirms Finnish sovereignty through modified physical presence in the islands, including airspace control, flight services, pilotage, coastguards and the role of the governor. Other such reminders are the adopted international treaties which are managed diplomatically by the Finnish Ministry for Foreign Affairs communicating with other states and with Ålandic authorities, as well as parliamentary debates and statements mentioned throughout this book.

22 Government of Åland, *Policy for Ålands demilitarisering och neutralisering—Handbok för landskapets myndigheter* (2015).

23 See e.g. the report Finnish Ministry for Foreign Affairs, *Åland och demilitarisering i dag, Utrikesministeriets publikationer* (1/2006).

5.3 The Åland Islands Regime as Part of Contemporary Collective Security

Technological developments have always posed new challenges for the law. From the introduction of rifles in the 19th century, to airplanes, telegraphy or nuclear technology in the 20th century, theoretical and principled debates on drones, digitalisation and cyberspace do not differ much today. They could, and still can, all be used for civil or military purposes, so there is inherent potential for dual usage. This is the case, for instance, with regard to space and surveillance technology. Furthermore, any such tool and any technology can be used for constructive or destructive, legal or illegal goals. Most wars have involved some form of new technology. This was the case in 1856, which saw the introduction of steamships and photography, to the First World War with the introduction of telegraphy, submarines and airplanes, through to modern surveillance radars and internet clouds.[24] Thus, technological development is not an obstacle *per se* to a legal and political institution such as the demilitarisation and neutralisation of the Åland Islands, nor is it in itself a guarantee that such an institution shall be functioning properly.[25] Furthermore, demilitarisation and neutralisation is not simply a military-strategic instrument; it is simultaneously a set of legal rules, a diplomatic and political tool and a confidence-building measure which transforms the relations between a number of different states and actors.[26]

These actors include the political institutions of the autonomous region (i.e. the Åland Government and Regional Parliament) as well as numerous international organisations with a certain role in the handling of the Åland regime, including the United Nations as depositary of the 1921 Åland Convention and as the main organisation responsible for collective security globally; the EU through the manifold recognitions and accommodation of the special international status of the Åland Islands; and the Nordic Council, where the Åland Islands hold a special position along with the Faroe Islands and Greenland.[27] The League of Nations

24 Spiliopoulou Åkermark, "Old Rules and New Technology: Drones and the Demilitarisation and Neutralisation of the Åland Islands" (Forthcoming).

25 For on overview of international law issues, including jurisdiction, the principle of distinction and the right to self-defence, see several chapters in N. Tsagourias and R. Buchan, *Research Handbook on International Law and Cyberspace* (2015). See also Ahlström, *Demilitarised and Neutralised Territories in Europe* (2004) 86–8.

26 S. Heinikoski, "*Ahvenanmaan asema ei ole vain puolustuspoliittinen kysymys*," *Politiikasta.fi*, 1 November 2016. Ahlström (2004).

27 S. Stephan, "*Självstyrelse och regional integration—Nordiska lösningar*," in *Självstyrelser i Norden i ett fredsperspektiv—Färöarna, Grönland och Åland*, eds. S. Spiliopoulou Åkermark and G. Herolf (2015) 70–87.

settlement guaranteed both the autonomous position of the islands and the protection of the Swedish language as well as the continued demilitarised and neutralised status entrenched in the Åland Convention. All this combined creates a complicated set of rights and obligations, expectations and assumptions surrounding the Åland Islands, at the domestic as well as international level. This explains the notion of the 'wooden puzzle', which has aptly been used by a seasoned Finnish diplomat, Ambassador René Nyberg, to describe the legal, political and geopolitical situation.[28]

Throughout this analysis and especially in Chapter 4 of this book, which deals with regional and sub-regional forms of security co-operation, the evolving concept of security has been discussed and grey zones and fusions between peace and war and between civil and military spheres described.[29] Such grey zones and fusions stretch valid constitutional and legal rules at the international and domestic levels and present us with difficult questions. How can the institution of demilitarisation and neutralisation be adapted to modern-day conditions? How should Finnish, and more generally domestic, as well as international law approach the use of force by non-state actors? How should new technology, such as drones or other unmanned vehicles on land and sea, be registered and regulated in order to ensure respect for the demilitarisation and neutralisation regime of the Åland Islands? Are rules of self-defence applicable in cyberspace or on satellites, and how does this affect the Åland Islands?[30] How can we implement democratic control and the rule of law in situations of complex emergencies where political and natural factors coincide and require multifaceted responses and action?[31]

28 R. Nyberg, "*Åland är som en träknut—lätt att plocka isär, men svår att sätta ihop,*" *Hufvudstadsbladet*, 24 November 2015.

29 The concept of a 'fusion' between civil and military spheres is not simply a construction for the purposes of this book. In 2008, NATO created the Civil-Military Fusion Centre, in Virginia (US) whose task was "[to] facilitate the sharing of open-source unclassified information between civilian and military actors working on complex crises in order to enhance their sense of shared awareness". It operated as such until 2013, after which its tasks were integrated into the work of the NATO Headquarters in Europe (Belgium) and its Comprehensive Crisis and Operations Management Centre (CCOMC).

30 Convention on Registration of Objects Launched Into Outer Space (1974) in Officials Records of the General Assembly, Twenty-Ninth Session, Supplement No. 20 (A/9620). While more than 90% of objects other than debris and non-functional objects in outer space are reported to be registered with the UN, in recent years the US has not been registering launched satellites.

31 The concept of 'complex emergencies' dates to the early 1990s when the UN General Assembly established the Inter-Agency Standing Committee (IASC) for the co-ordination of humanitarian assistance, through Resolution 46/182 (19 December 1991). In 1994, the IASC defined "complex emergency" as a humanitarian crisis involving a "breakdown of

What does "a move from heteronomy", in Grimm's sense (see section 5.2), mean in today's globalised world and how does that affect an autonomous and demilitarised region such as the Åland Islands? Has the idea of sovereignty really lost much of its normative or descriptive meaning, or does it allow for the creation of political spaces for new debates about competing rationalities and goals and thus for empowerment and liberty?

Technology presents, however, not only difficulties and threats. It also offers opportunities. Modern communications technology means, for instance, that people on the Åland Islands have broader channels of information and contact than was the case a century ago, when the newspaper *Ålandstidningen* was the only main source of broader information. Such technology allows for territorial surveillance of the demilitarised region from outside the demilitarised zone itself and allows for co-operation, such as surveillance in the Baltic Sea region, first between Finland and Sweden and later including several other countries, as we have seen with regard to SUCFIS and SUCBAS (see section 4.3).[32] All such developments have the potential of reaffirming and strengthening the demilitarisation and neutralisation of the Åland Islands, as well as of undermining it.

5.4 Afterword: Does the Demilitarisation Add Anything to the Quest for Peace in the Baltic Sea?

One of the paradoxes of demilitarisation concerns the continuum of militarisation–demilitarisation. While the demilitarised region experiences a decline or abolition of military presence, there is the need for the state (or possibly some other international actor and guarantor in other cases) to increase its intelligence, border control and military capability, in order to be able to respond to challenges along the borders of the demilitarised region within the margins authorised explicitly by the 1921 Convention. In our case study, covering many decades, the management and responses have been mainly diplomatic and legal (negotiating and managing treaties, increasing information and transparency in the system, clarifying and updating borders, expanding the territorial sea, communication between Finnish and Ålandic authorities in order to solve possible problems etc.). However, the primacy of diplomacy and law is being questioned today, with more attention paid to military tools.

Militarisation in the political sense can be observed in increased attention being directed towards military affairs, a process which often involves

---

authority" as a result of conflict or foreign aggression, as well as necessitating the involvement of several international actors and thereby their co-ordination.

32 A. Gardberg, *Åland Islands: A Strategic Survey* (1995).

securitisation of issues that were not previously considered security matters. It is also important to trace the processes of securitisation and militarisation; 'securitisation' as a term was coined in the 1990s, when the phenomenon of making security was conceptualised.[33] Along with securitisation, we argue that militarisation has occurred as a result of the threat-focused securitisation discourses. After the end of the Cold War, the use of force seemed to be less urgent but also freer. The Balkan conflict brought war in Europe, and the EU started to progressively outline its Common Foreign and Security Policy (CFSP) established in 1993, and the European Security and Defence Policy (ESDP), established in 1999. Many European states wanted to integrate the military alliance WEU into the EU, something which effectively occurred in 2009 with the entry into force of the Lisbon Treaty. Both at the rhetorical and institutional level, we see trends of militarisation and decreasing parliamentary involvement in matters concerning security and defence matters. Many defence co-operation arrangements are made in the form of non-legally binding Memoranda of Understanding and are not seen as having to be subjected to parliamentary approval.

Regionalism can take the form of military co-operation, and then one could possibly have the formation of a 'regional identity' based primarily on security and military priorities. Increased military spending and a strengthening of the political weight of military issues and of the military are likely outcomes. However, as argued in Chapter 1, the demilitarised and neutralised status of the Åland Islands rests upon multipolarity and does not rely simply upon strong international institutions, or on Cold War bipolarity, or on global hegemony. Not surprisingly, the role of neighbouring states and great powers has been crucial in ensuring the survival of the regime, but such international actors are joined in this effort by national actors such as the Finnish Parliament and Government, the President of the Republic and an active approach adopted early on by the Ålanders, including Ålandic elites and institutions.

A further paradoxical situation of the demilitarisation concerns the secrecy-transparency perspective. While military affairs are understood as requiring secrecy, especially at times of tension and conflict, in order to, as is argued, enhance deterrence and minimise risks for military takeover, the demilitarisation 'logic' by contrast assumes a considerable measure of openness. The demilitarised region needs not only to be formally demilitarised, but it needs also to appear demilitarised, so to say. This is confirmed in the present study of the Åland case, including the role of the Swedish and Russian

33 O. Waever, "Securitization and Desecuritization," in *On Security*, ed. R.D. Lipschutz (1995) 46–86.

consulates, the role of the Governor in managing the treaties on Åland, the establishment of a Contact Group between the Finnish Ministry for Foreign Affairs and the Åland Government and annual meetings between the Finnish Coast Guard leadership and the Åland Government as well as between the Finnish Marine and said government. The press releases of the Finnish Coast Guard and Ministry of Defence, which started being issued in 2005, concerning alleged violations of the demilitarised zone, are another example of this effort towards openness.[34] This is an additional reason why secret surveillance co-operation activities affecting the Åland Islands pose considerable problems.

The demilitarisation and neutralisation of the Åland islands is a relational concept which has profoundly and consistently modified the status of these islands at the local, national and international levels. It is not a radical pacifistic tool which precludes all kinds of military activities in the archipelago, but rather a pragmatic tool of diplomatic co-operation and constraint in the use of force. The rules on the demilitarisation and neutralisation of the islands restrict access to the islands first and foremost for military forces of third countries, but also to a more limited extent for the Finnish military, without, however, extinguishing the right to self-defence nor the obligation of Finland to uphold the demilitarised zone. This effort is to be pursued foremost by diplomatic and legal means, following both the demilitarisation and neutralisation undertakings as well as the pillars of collective security under the UN Charter. The Åland regime, with its elements of demilitarisation and neutralisation, autonomous status and cultural guarantees—but under Finnish sovereignty and constitutional order—forms part of the idea of a rules-based international society and the rule of law in international affairs, something which is usually considered as an asset for small states and weaker actors.

Åland has not been as devastated by war and conflict as one might have expected, taking into account recurring references to its geostrategic importance.[35] There has been some military presence during the wars, but such instances have either been kept within the limits and margins of the legal regime, or have been accepted by the other states parties. While the legal validity of the demilitarisation and neutralisation regime has been strengthened over its more than 160 years of existence, the relative political and geopolitical importance of the islands shall continue to vary depending on the power balances and historical and political contingencies, domestic as well as international, at each point of time. The demilitarisation and

34 Information received from The Finnish Border Guard, letter dated 17 May 2017. On file with the authors.

35 Björkholm and Rosas (1990) 15–16.

neutralisation of the Åland Islands remains one of the tools for maintaining peace in the Baltic Sea, alongside other and ongoing efforts to limit the use of armed force globally and regionally, including discussions on the limitation and banning of nuclear weapons. As we have tried to show throughout this book, the demilitarisation and neutralisation of the Åland Islands makes sense only if seen as a small part of the global collective security regime. It has been one of the tools available for enhancing rules-based continuity as well as flexibility in Finnish foreign policy. It remains to be seen how the integrity and validity of the demilitarisation regime shall be upheld at times of enhanced civil-military co-operation and a trend towards increased use of armed force alongside a fusion between war and peace.

Appendix

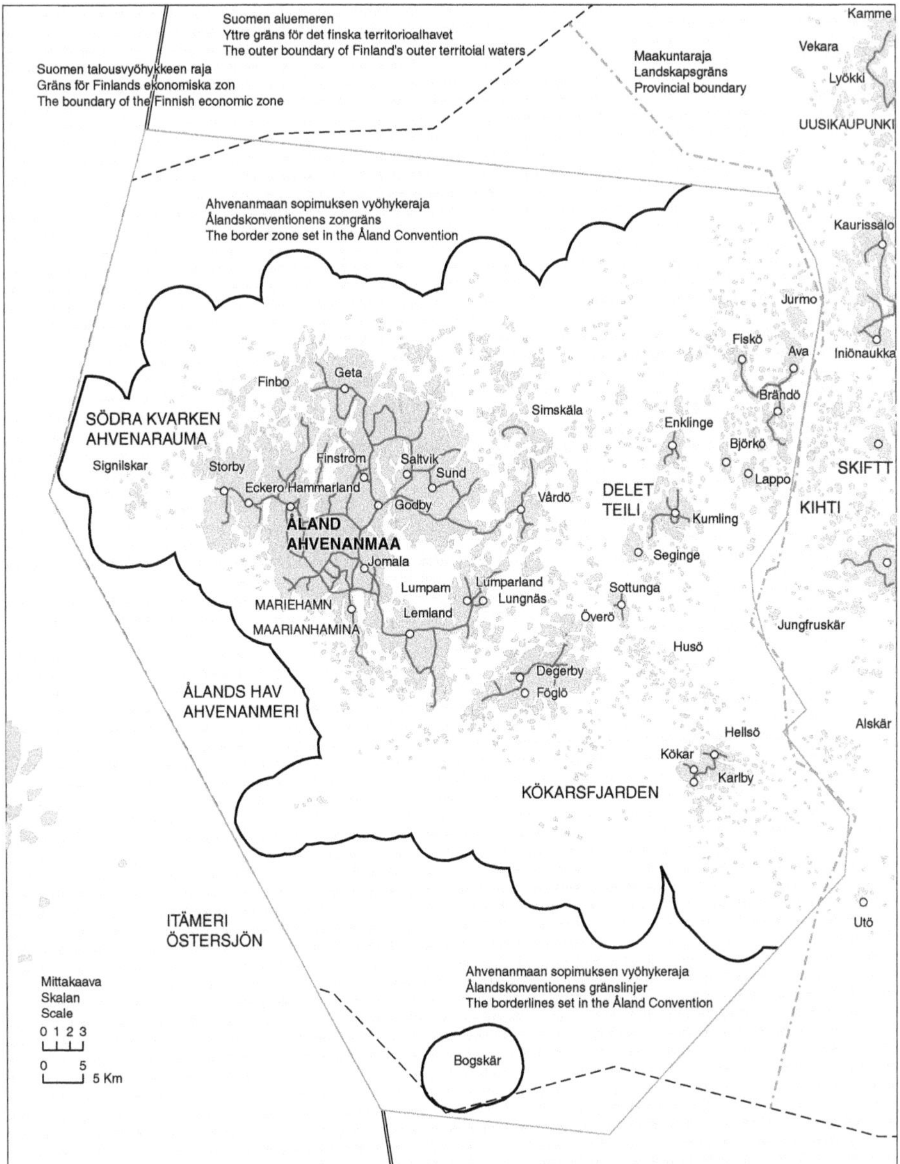

Figure A.1 Boundaries of the demilitarised and neutralised zone of the Åland Islands.

Source: Map produced by the National Land Survey and Ministry for Foreign Affairs of Finland (2013), available at formin.finland.fi/public/default.aspx?contentId=272970&nodeId=49150&contentlan=2&culture=en-US

Bibliography

Abulafia, David. "Mediterranean History as Global History." *History and Theory* 50, No. 2 (2011) 220–28.

Aggestam, Lisbeth. "Introduction: Ethical Power Europe?" *International Affairs* 84, No. 1 (2008) 1–11.

Agyebeng, William K. "Theory in Search of Practice: The Right of Innocent Passage in the Territorial Sea." *Cornell International Law Journal* 39 (2006) 371–89.

Ahlström, Christer. *Demilitarised and Neutralised Territories in Europe*. Mariehamn: Ålands fredsinstitut, 2004.

Akaha, Tsuneo. "Japan's Comprehensive Security Policy: A New East Asian Environment." *Asian Survey* 31, No. 4 (1991) 324–40.

Åkerström, David. *Nätverksbaserat försvar—Då och Nu*. Stockholm: Försvarshögskolan, 2012.

Ålands kulturstiftelse. *Internationella avtal och dokument rörande Åland 1856–1992*. Mariehamn: Ålands kulturstiftelse, 1993.

Ålands lagting. "*Självstyrelsepolitiska nämnden*." Förslag nr 1/2001–2002.

Ålandstidningen. "*Ålänningarna och värnplikten* [The Ålanders and Military Service]." 14 January 1920.

Ålandstidningen. 1 November 1938, 3 November 1938 and 20 August 2010 (re. The Farmers' March).

Anand, Ram Prakash. "Transit Passage and Overflight in International Straits." *Indian Journal of International Law* 26 (1986) 72–105.

"The Atlantic Conference: Joint Statement by President Roosevelt and Prime Minister Churchill, 14 August 1941." 1941.

Australian International Law News (1990), 77. "Secret Protocols to the Molotov–Ribbentrop Pact".

Bailes, Alyson J.K. "Parliaments and National Strategy Documents: A Comparative Case-Study from the Nordic Region." Geneva Centre for the Democratic Control of Armed Forces (DCAF), *Policy Paper No. 36*, 2015.

Bakker, Anne, Sven Biscop, Margriet Drent, and Lennart Landman. "Spearheading European Defence Employing the Lisbon Treaty for a Stronger CSDP." Netherlands Institute of International Relations, *Clingendael Report*, 2016.

Bangert, Kaare. "Internal Waters." *Max Planck Encyclopedia of Public International Law*.

Barnett, Robert W. *Beyond War: Japan's Concept of Comprehensive National Security*. Washington: Pergamon-Brassey's, 1984.

Barros, James. *The Aland Islands Question: Its Settlement by the League of Nations*. New Haven: Yale University Press, 1968.

Bauwens, Werner, Nicholas Sherwen, Armand Cless, and Olav F. Knudsen. *Small States and the Security Challenge in the New Europe*. 1st English edn. Vol. 8. London: Brassey's, 1996 (1943).

Beck, Ulrich. *World Risk Society*. Cambridge: Polity Press, 1999.

Bederman, David J. "Collective Security, Demilitarization and 'Pariah' States." *European Journal of International Law* 13, No. 1 (2002) 121–38.

Bellamy, Alex J. *Security Communities and Their Neighbours: Regional Fortresses or Global Integrators*? Basingstoke: Palgrave Macmillan, 2004.

Ben Slimane, Fatma. "Between Empire and Nation-State: The Problem of Borders in the Maghreb." In *Mediterranean Frontiers: Borders, Conflict and Memory in a Transnational Word*, edited by Kalypso Nicolaidis and Dimitar Bechev. London: Tauris, 2009, 35–55.

Berghahn, Volker R. *Militarism: The History of an International Debate 1861–1979*. Leamington Spa: Berg/Cambridge University Press, 1981.

Beyens baron, Eugène-Napoléon. *L'Allemagne avant la guerre*. Oxford: Université d'Oxford, 1915.

Bickford, Andrew. "Demilitarization: Unraveling the Structures of Violence." In *Demilitarization in the Contemporary World*, edited by Peter N. Stearns. Urbana: University of Illinois Press, 2013, 21–34.

Biscop, Sven. "All or Nothing? The EU Global Strategy and Defence Policy After the Brexit." *Contemporary Security Policy* 37, No. 3 (2016) 431–45.

Björkholm, Mikaela, and Allan Rosas. *Ålandsöarnas demilitarisering och neutralisering*. Åbo: Åbo Akademis förlag, 1990.

Blake, R.L.V.F. *The Crimean War*. Barnsley: Pen and Sword Books, 1970, reprinted 2006.

Bodin, Jean. *Les six livres de la république* (first published 1576).

Bondestam, Anna. *Åland Vintern 1918*. Vol. 6. Mariehamn: Ålands kulturstiftelse, 1972.

Bönker, Dirk. *Militarism in a Global Age: Naval Ambitions in Germany and the United States Before World War I*. Ithaca: Cornell University Press, 2012.

Bringéus, Krister. *"Säkerhet i ny tid", SOU 2016:57*. Stockholm: Statens Offentliga Utredningar, 2016.

Brown, Edward Duncan. *The International Law of the Sea, Volume Introductory Manual*. Dartmouth: Dartmouth Publishing Company Limited, 1994.

Browning, Christopher S. "Towards a New Agenda? US, Russian and EU Policies in Northern Europe." In *NEBI Yearbook 2003*, edited by Lars Hedegaard and Bjarne Lindström. New York: Springer, 2003, 273–89.

Browning, Christopher S., and Pertti Joenniemi. "Regionality Beyond Security? The Baltic Sea Region After Enlargement." *Cooperation and Conflict* 39, No. 3 (2004) 233–53.

Brownlie, Ian. *Principles of Public International Law*. 6th edn. Oxford: Oxford University Press, 2003.

Buzan, Barry. *The European Security Order Recast: Scenarios for the Post-Cold War Era*. London: Pinter Publishers, 1990.
Buzan, Barry, Ole Waever, and Jaap de Wilde. *Security: A New Framework for Analysis*. Boulder and London: Lynne Rienner Publishers, 1998.
Caminos, Hugo, and Michael R. Molitor. "Progressive Development of International Law and the Package Deal." *American Journal of International Law* 79, No. 4 (1985) 871–90.
Carr, Edward Hallett. *A History of Soviet Russia: The Bolshevik Revolution 1917–1923*. Vols. 1–3. London: Macmillan, 1985.
Carrapico, Helena. "Analysing the European Union's Responses to Organized Crime Through Different Securitization Lenses." *European Security* 23, No. 4 (2014) 601–17.
Castellino, Joshua, Steve Allen, and Jérémie Gilbert. *Title to Territory in International Law: A Temporal Analysis*. Aldershot: Ashgate, 2003.
Centre de Droit International. "A Plea Against the Abusive Invocation of Self-Defence as a Response to Terrorism." *Université Libre de Bruxelles*.
Churchill, R.R., and A.V. Lowe. *The Law of the Sea*. 3rd edn. Manchester: Manchester University Press, 1999.
Commission Internationale de Juristes. "Ålandsfrågan inför Nationernas Förbund: La Question des Îles d'Aland. Rapport de la Commission Internationale de Juristes (5 September 1920)." In *Diplomatiska Aktstycken Utgivna Av Kungliga Utrikesdepartementet*. Stockholm: Utrikesdepartementet, 1920.
Conference of the Representatives of the Governments of the Member States. "Cover Note CIG 62/03." 5 December 2003.
"Confidential Note." Date 23 July 1981, signed F.D. Berman, National Archives, FCO 033/5081.
"Consolidated Versions of the Treaty on European Union and the Treaty on the Functioning of the European Union." *Official Journal of the European Union*, 2016.
Constantinou, Costas M., and Sam Okoth Opondo. "Engaging the 'Ungoverned': The Merging of Diplomacy, Defence and Development." *Cooperation and Conflict* 51, No. 3 (2016) 307–24.
Constitution of the Republic of Finland, 731/1999 (in force 2000).
Convention on Registration of Objects Launched into Outer Space (1974). *Officials Records of the General Assembly, Twenty-Ninth Session*, Supplement No. 20 (A/9620).
Council of the European Union. "2298th Meeting of the Permanent Representatives Committee held in Brussels on 2–4 and 7 December 2009." Brussels, 2010.
——— "Implementation Plan on Security and Defence." 14 November 2016.
Dalby, Simon. "Realism and Geopolitics." In *The Ashgate Research Companion to Critical Geopolitics*, edited by Klaus Dodds, Merje Kuus, and Joanne Sharp. Farnham: Ashgate, 2013, 33–47.
Delbrück, Jost. "Max Huber's Sociological Approach to International Law Revisited." *European Journal of International Law* 18, No. 1 (2007) 97–113.
Devine, Karen. "Neutrality and the Development of the European Union's Common Security and Defence Policy: Compatible or Competing?" *Cooperation and Conflict* 46, No. 3 (2011) 334–69.

Duchêne, Francois. "Europe's Role in World Peace." In *Europe Tomorrow: Sixteen Europeans Look Ahead*, edited by Richard Mayne. London: Fontana for Chatham House and P.E.P, 1972, 32–47.

Dyzenhaus, David. *The Constitution of Law: Legality in a Time of Emergency*. Cambridge: Cambridge University Press, 2006.

Elden, Stuart. *The Birth of Territory*. Chicago: The University of Chicago Press, 2013.

Erich, Rafael. *Suomen valtio-oikeus I osa*. Helsinki: WSOY, 1924.

European Council. "European Council Meeting (20–21 March 2014)—Conclusions." 21 March 2014.

——— "European Council Meeting (22–23 June 2017)—Conclusions." 23 June 2017.

European Parliament. "European Parliament Resolution of 22 November 2016 on the European Defence Union (2016/2052(INI))." 22 November 2016.

Fagerlund, Niklas. *Innebörden av uttrycket temporärt förankra i de till öarna hörande vatten i artikel 4 st. 2b) Ålandskonventionen*. Mariehamn: Åländsk utredningsserie, 1994.

——— "The Special Status of Åland Islands in the European Union." In *Autonomy and Demilitarisation in International Law: The Åland Islands in Changing Europe*, edited by Lauri Hannikainen and Frank Horn. The Hague and Cambridge: Kluwer Law International, 1997, 189–256.

Federal Government of Germany. "White Paper on German Security Policy and the Future of the Bundeswehr." 13 July 2016 (J.-M. Ayrault and F.-W. Steinmeier).

Finnish Act on the Autonomy of Åland (1144/1991).

Finnish Act on the Decision-making Concerning Provision and Reception of International Assistance (418/2017).

Finnish Government. "*Perustuslain 97 §:n mukainen selvitys eduskunnan ulkoasiainvaliokunnalle Suomen kansainvälisestä sopimuspolitiikasta ja sen kehityssuunnista* [Report Under Section 97 of Constitution for the Committee for Foreign Affairs of the Parliament on Finnish International Agreement Policy and Its Development]." 2011.

——— "*EU-ministerivaliokunnan kokous linjasi kantoja Eurooppa-neuvostoon* [Meeting of the Ministerial Committee on European Union Affairs Formulated Stances for European Council Meeting]." 20 June 2017.

——— Bill 94/2016 on Proposed Changes of Legislation Pertaining to the Act on Armed Forces, the Territorial Surveillance Act and the Act on Conscription.

——— Bill 72/2016 Concerning Provision and Acceptance of International Assistance.

——— Bill 107/2016 on Assistance in Matters Falling Under the Competence of the Ministry of Interior.

Finnish Ministry of Defence. "Press Release No. 65–2003." Nordic Peace 2003—*Harjoitus ja Ahvenanmaa*, 12 September 2003.

Finnish Ministry for Foreign Affairs. "Åland och demilitarisering i dag, Utrikesministeriets publikationer." 1/2006.

——— "*Ålands demilitarisering och neutralisering mot bakgrund av de nya europeiska utmaningarna* [Åland Demiltarisation and Neutralisation against the New European Challenges]." 2010.

——— *The Effects of Finland's Possible NATO Membership: An Assessment*. Edited by Mats Bergquist, François Heisbourg, René Nyberg, and Teija Tiilikainen. Finnish Ministry for Foreign Affairs, 2016.
Finnish Ministry of Justice. "*En utredning om gränserna för Ålands demilitarisering*." 2006.
——— "Final Report of the Committee for a Revision of the Self-Government Act for Åland." Report 33(2017): *Ålands självstyrelse i utveckling*. 2017.
Finnish Parliament. "Minutes of the Plenary Session on 31 June 1995." 1995.
——— "Minutes of the Plenary Session on 31 October 1995." 1995.
——— "Minutes of the Plenary Session on 1 November 1995." 1995.
——— "Minutes of the Plenary Session 28/1997 on 17 March 1997." 1997.
——— "Minutes of the Plenary Session on 18 March 1997." 1997.
——— "Minutes of the Plenary Session on 10 December 2001." 2001.
——— "Written Question 457/2003 vp: Compliance with the Agreement Arrangements Concerning the Demilitarisation of the Åland Islands." 12 November 2003.
——— "Minutes of the Plenary Session on 22 June 2004." 2004.
——— "Minutes of the Plenary Session on 28 September 2004." 2004.
——— "Minutes of the Plenary Session on 29 September 2004." 2004.
——— "Minutes of the Plenary Session on 20 December 2004." 2004.
——— "Minutes of the Plenary Session on 14 September 2006." 2006.
——— "Minutes of the Plenary Session on 10 October 2006." 2006.
——— "Minutes of the Plenary Session on 22 November 2006." 2006.
——— "Minutes of the Plenary Session on 11 December 2014." 2014.
——— "Minutes of the Plenary Session on 21 June 2016." 2016.
——— "Minutes of the Plenary Session on 20 October 2016: Oral Question on the Demilitarisation of the Åland Islands." 2016.
——— "Minutes of the Plenary Session on 21 December 2016." 2016.
——— "Minutes of the Plenary Session on 8 March 2017." 2017.
Finnish Parliament Administration Committee. "Foreign Affairs Committee Report on the Government Report 1/1995." 1995.
——— "Statement of the Committee for Foreign Affairs 4/2004 Government Report on Security and Defence Policy 2004." 2004.
——— "Statement on the Government Report on Foreign and Security Policy HaVL 40/2016." 2016.
——— "Statement of the Foreign Affairs Committee on the Government Defence Report." 2017.
Finnish Parliament Constitutional Committee. "Statement of the Committee on Constitutional Law on Government Bill 135/1994." 1994.
Finnish Parliament Defence Committee. "Statement of the Defence Committee on the Government Report 1/1995." 1995.
——— "Report of the Defence Committee on the Government Security and Defence Policy Report." 2004.
——— "Report of the Defence Committee on the Government Defence Report." 2017.
——— "Opinion 9/2009 Concerning EU Commission Proposal on the Integration of Maritime Surveillance in COM(2009)538 Final".

Finnish Treaty Series 1/1922. "Convention Relating to the Non-Fortification and Neutralisation of the Åland Islands (1921)." English translation available in 17 AJIL 1923, Supplement: Official Documents.

Finnish Treaty Series 24/1940. "Treaty Concerning the Åland Islands Between Finland and the Union of Soviet Socialist Republics".

Finnish Treaty Series 9/1948. "Reactivation of the Previous Treaty Between Finland and the Union of Soviet Socialist Republics".

Finnish Treaty Series 20/1947. "Peace Treaty With Finland." English translation available in 42 AJIL 1948, Supplement: Official Documents.

Finnish Treaty Series 63/1992. "*Överenskommelse Med Ryska Federationen Om Grunderna För Relationen Mellan Länderna.*" 1992.

"The First Conference for the Codification of International Law." *American Journal of International Law* 41(1947), Supplement.

Friis, Karsten, and Erik Rechborn-Kjennerud. "From Cyber Threats to Cyber Risks." In *Conflict in Cyber Space: Theoretical, Strategic and Legal Perspectives*, edited by Karsten Friis and Jens Ringsmose. Abingdon, Oxon: Routledge, 2016, 27–44.

Froman, F. David. "Uncharted Waters: Non-innocent Passage of Warships in the Territorial Sea." *San Diego Law Review* 21 (1984) 625–89.

Gardberg, Anders. *Åland Islands: A Strategic Survey*. Helsinki: National Defence University, 1995.

Gavouneli, Maria. "Neutrality—A Survivor?" *European Journal of International Law* 23, No. 1 (2012) 267–73.

Gebhard, Carmen. "Soft Competition: Finland, Sweden and the Northern Dimension of the European Union." *Scandinavian Political Studies* 36, No. 4 (2013) 365–90.

Gestrin-Hagner, Maria. "Huvudstaben förbjöd ryskt skolfartyg att besöka Åland." *Hufvudstadsbladet*, 28 August 2017.

Goodhart, Michael. "Europe's Democratic Deficits Through the Looking Glass: The European Union as a Challenge for Democracy." *Perspectives on Politics* 5, No. 3 (2007) 567–84.

Götz, Norbert. "The Case of the Baltic Sea Area: Spatial Politics & Fuzzy Regionalism." *Baltic Worlds* 9, No. 3 (2016) 54–67.

Government of Åland. "*Policy för Ålands demilitarisering och neutralisering* [Policy for Åland's Demilitarisation and Neutralisation]." In *Handbok för landskapets myndigheter*. Åland: Government of Åland, 2015.

Gregory, Charles Noble. "The Neutralization of the Aaland Islands." *American Journal of International Law* 17 (1923) 63–76.

Grimm, Dieter. *Sovereignty: The Origin and Future of a Political and Legal Concept*. New York: Columbia University Press, 2015.

Grotius, Hugo. *Jure Belli Ac Pacis Libri Tres* [On the Law of War and Peace: Three Books]. Oxford: Clarendon Press, 1646/1925.

Gummesson, Jonas. "SvD/Sifo: Kraftigt ökat motstånd mot Nato." *Svenska Dagbladet*, 2 July 2016.

Gustavsson, Kenneth. *Ålandsöarna—en säkerhetsrisk?: spelet om den demilitariserade zonen 1919–1939*. Mariehamn: PQR-kultur, 2012.

Gutteridge, Joyce Ada Cooke. *The United Nations in a Changing World*. Manchester: Manchester University Press, 1969.

Haglund, Carl. "The Baltic Sea as an Example of Regional Maritime Security Cooperation." *Baltic Rim Economies Review*, No. 4 (2014) 4.

Hakapää, Kari. *Uusi kansainvälinen oikeus*. Hämeenlinna: Talentum, 2010.

Halldenius, Lena. *Liberty Revisited: A Historical and Systematic Account of an Egalitarian Conception of Liberty and Legitimacy*. Lund: Bokbox Publications, 2001.

Hannikainen, Lauri. "The Continued Validity of the Demilitarised and Neutralised Status of the Åland Islands." *Zeitschrift Für Ausländisches Öffentliches Recht Und Völkerrecht* 54, No. 3 (1994) 614–51.

Hannikainen, Lauri, and Frank Horn. *Autonomy and Demilitarisation in International Law: The Åland Islands in a Changing Europe*. Edited by Lauri Hannikainen and Frank Horn. The Hague: Kluwer Law International, 1997.

Heinikoski, Saila. "Ahvenanmaan asema ei ole vain puolustuspoliittinen kysymys." *Politiikasta.fi*, 1 November 2016.

——— "The Åland Islands, Finland and European Security in the 21st Century." *Journal of Autonomy and Security Studies* 1, No. 1 (2017) 8–45.

——— " 'Pool It or Lose It'—Discourses on EU Military Integration and Demilitarisation in the Baltic Sea." *Journal on Baltic Security* 3, No. 1 (2017) 32–47.

——— "Path Dependency and Foreign Policy Change—Finnish Discourses on Post-Neutrality and Demilitarisation After EU Accession" (forthcoming).

Heintschel von Heinegg, Wolff. "Warships." *Max Planck Encyclopedia of Public International Law*, 2015.

Helsinki Commission (HELCOM). "Ensuring Safe Shipping in the Baltic." 2009.

Herolf, Gunilla. "Cooperation in the North—Multilateralism or Mess?" *Mercury E-paper No. 7*, 2010, 1–11.

Hobbes, Thomas. *Leviathan* (first published 1651).

Hoogensen Gjørv, Gunhild. *Understanding Civil-Military Interaction*. Surrey: Ashgate, 2014.

Howorth, Jolyon. "Why ESDP Is Necessary and Beneficial for the Alliance." In *Defending Europe: The EU, NATO, and the Quest for European Autonomy*, edited by Jolyon Howorth and John T.S. Keeler. Gordonsville: Palgrave Macmillan, 2003, 219–38.

Hübner, Emanuel. "Some Notes on the Preparations for the Olympic Games of 1936 and 1940: An Unknown Chapter in German-Finnish Cooperation." *The International Journal of the History of Sport* 30, No. 9 (2013) 950–62.

Hughes, C.W. "Japan's Remilitarization and Constitutional Revision." In *Demilitarization in the Contemporary World*, edited by Peter N. Stearms. Urbana: University of Illinois Press, 2013, 127–156.

Innerarity, Daniel. *The Transformation of Politics: Governing in the Age of Complex Societies*. Brussels and New York: Peter Lang, 2010.

Isaksson, Martin. *Ryska positionen Alandskaja. En översikt av Ålands militära historia åren 1906–1918*. Ekenäs: Söderström & Co, 1983.

Jakobson, Max. *Finland in the New Europe*. Vol. 175. Westport: Praeger with the Center for Strategic and International Studies, 1998.

Jellinek, Georg. *Allgemeine Staatslehre*. 3rd edn. Berlin: O. Häring, 1914.

Jessup, Philip C. *The Law of Territorial Waters and Maritime Jurisdiction 24*. Vol. 24. New York: G.A. Jennings Co., Inc., 1927.

Joenniemi, Pertti. "The EU Strategy for the Baltic Sea Region: A Catalyst for What?" *DIIS Brief*, August 2009.

Jyränki, Antero, Jaakko Husa, and Sten (transl. Palmgren). *Konstitutionell rätt*. Helsingfors: Talentum, 2015.

Kaukoranta, Päivi. "Negotiation of the Treaty on Open Skies Revisited: Finnish Features." In *Nordic Cosmopolitanism: Essays in International Law for Martti Koskenniemi*, edited by Jarna Petman and Jan Klabbers. Leiden: Martinus Nijhoff Publishers, 2003, 371–89.

Kingma, Kees, and Nico Schrijver. "Demilitarization." In *Max Planck Encyclopedia of Public International Law*. Oxford: Oxford University Press, 2013.

Kleemola-Juntunen, Pirjo. *Passage Rights in International Law: A Case Study of the Territorial Waters of the Åland Islands*. Rovaniemi: University of Lapland, 2014.

——— "The Right of Innocent Passage: The Challenge of the Proliferation Security Initiative and the Implications for the Territorial Waters of the Åland Islands." In *The Future of the Law of the Sea: Bridging Gaps Between National, Individual and Common Interests*, edited by Gemma Andreone. Cham: Springer, 2017, 239–69.

——— "Straits in the Baltic Sea: What Passage Rights Apply?" In *Regulatory Gaps—Selected Issues in Baltic Sea Governance*, edited by Henrik Ringbom (forthcoming 2018).

Klein, Natalie. *Dispute Settlement in the UN Convention on the Law of the Sea*. Cambridge: Cambridge University Press, 2009.

Knop, Karen, and Susan Marks. "The War Against Cliché: Dispatches From the International Legal Front." In *Sovereignty, Statehood and State Responsibility: Essays in Honour of James Crawford*, edited by Christine Chinkin and Freya Baetens. Cambridge: Cambridge University Press, 2015, 3–21.

Koivisto, Mauno. *Witness to History: The Memoirs of Mauno Koivisto, President of Finland 1982–1994*. London: Hurst, 1997.

Koivurova, Timo, and Filip Holiencin. "Demilitarisation and Neutralisation of Svalbard: How Has the Svalbard Regime Been Able to Meet the Changing Security Realities During Almost 100 Years of Existence?" *The Polar Record* 53, No. 2 (2017) 131–42.

Kontio, Pirjo. "Juha Sipilä: Nato-jäsenyys rajoittaisi Suomen liikkumatilaa [NATO Membership Would Restrict Finland's Freedom to Manoeuvre]." *Suomenmaa*, 15 June 2014.

Koskenniemi, Martti. *The Gentle Civilizer of Nations: The Rise and Fall of International Law, 1870–1960*. Cambridge, UK and New York: Cambridge University Press, 2001.

——— "Occupation and Sovereignty—Still a Useful Distinction?" In *Law at War—The Law as It Was and the Law as It Should Be*, edited by Ola Engdahl and Pål Wrange. Leiden: Brill, 2008, 163–74.

——— "What Use for Sovereignty Today?" *Asian Journal of International Law* 1, No. 1 (2011) 61–70.

——— "Diplomats, Professors, and Then Some: Notes for a History of International Law in 20th Century Finland." *Nordic Journal of International Law* 85, No. 4 (2016) 322–33.

Lagoni, Rainer. "Internal Waters", 11 *Encyclopedia of Public International Law*, 1989, 153–55.

Lambert, Andrew. "Bomarsund i ett internationellt perspektiv. En fästningsuppgång och fall och dess plats i den globala konkurrensens strategiska kultur." *Åländsk odling 2004–2005*, 152–187.

Lavery, Jason. "Finnish-German Submarine Cooperation 1923–35." *Scandinavian Studies* 71, No. 4 (1999) 393–418.

Lax, Henrik. "The Åland Regime After Crimea." *Baltic Rim Economies Review*, No. 5 (2015) 10.

League of Nations. "General Treaty for Renunciation of War as an Instrument of National Policy." *League of Nations, Treaty Series*, 1929, Vol. XCIV.

———. "Official Journal." May–June 1939.

——— "Report on the Work of the League 1938/39." August 12th, 1939, Series of League of Nations Publications General 1939.

Lehmann, Jean-Pierre. "Book Review: Japan's Quest for Comprehensive Security: Defence, Diplomacy and Dependence (J.W.M. Chapman, Reinhard Drifte, I.T.M. Gow)." *International Affairs (Royal Institute of International Affairs 1944–)* 60, No. 1 (1983–1984).

"Letter of 4 September 1939 From the Ministry of Interior to the Governor on Åland Concerning the Implementation of Relevant Legislation." *Länsstyrelsens arkiv*, Ea42a, Åland Archives.

Leuprecht, Peter. *Reason, Justice and Dignity: A Journey to Some Unexplored Sources of Human Rights*. Boston and Leiden: Martinus Nijhoff Publishers, 2011.

Linderfalk, Ulf. "International Legal Hierarchy Revisited—The Status of Obligations *Erga Omnes*." *Nordic Journal of International Law* 80, No. 1 (2011) 1–23.

Lundgren, Bengt. "Security and Surveillance Cooperation in the Baltic." Slides with notes.

Malanczuk, Peter, and Michael Barton Akehurst. *Akehurst's Modern Introduction to International Law*. London: Routledge, 2002.

Manners, Ian. "Normative Power Europe: A Contradiction in Terms?" *Journal of Common Market Studies* 40, No. 2 (2002) 235–58.

——— "Normative Power Europe Reconsidered: Beyond the Crossroads." *Journal of European Public Policy* 13, No. 2 (2006) 189–99.

"Memorandum of Understanding on the European Centre of Excellence for Countering Hybrid Threats." Helsinki, 11 April 2017.

"Minutes dated 20 November 1951." Signed by Joyce Gutteridge. National Archives, FCO 033/5081.

Moberg, Mikael, James Mashiri, and Charly Salonius-Pasternak. "Vaara! vihreitä miehiä Ahvenanmaalla!—Näin se voisi tapahtua." *Suomen Kuvalehti*, 2 August 2015.

Modeen, Tore. *De folkrättsliga garantierna för bevarandet av Ålandsöarnas nationella karaktär*. Åbo; Helsingfors & Mariehamn: Finlands juristförbund, 1973.

The New York Times. "Which Countries Support and Which Oppose the U.S. Missile Strikes in Syria." 9 April 2017.

Ngantcha, Francis. *The Right of Innocent Passage and the Evolution of the International Law of the Sea: The Current Regime of Free Navigation in Coastal Waters of Third States*. Geneva: Pinter Publishers, 1990.

Niinistö, Jussi. "Ahvenanmaan asia." *Jussi Niinistö's Blog*, 17 October 2016.

——— "A demilitarised Åland is a military vacuum." *Helsinki Times*, 18 October 2016.

——— "Itämeren geostrateginen merkitys kasvussa." *Pulloposti*, No. 9 (2017).

Noortmann, Math. *Enforcing International Law: From Self-Help to Self-Contained Regimes*. Abingdon: Routledge, 2016.

Nordquist, Myron H. et al. *United Nations Convention on the Law of the Sea 1982*. Vol. V. Leiden: Martinus Nijhoff Publishers, 1989.

——— (eds.). *United Nations Convention on the Law of the Sea 1982: A Commentary*. Vol. II. Leiden: Martinus Nijhoff Publishers, 1993.

Nyberg, René. "Finlands säkerhetspolitik och Ålands status." *Hufvudstadsbladet*, 26 September 1990.

——— "Ni har vidrört VSB-avtalet." In *Säkerhetspolitik och historia*, edited by Krister Wahlbäck, Mats Bergquist, and Alf W. Johansson. Stockholm: Hjalmarson & Högberg, 2007, 285–99.

——— "Åland är som en träknut—lätt att plocka isär, men svår att sätta ihop." *Hufvudstadsbladet*, 24 November 2015.

OCHA. "Civil-Military Guidelines and Reference for Complex Emergencies." *United Nations*, 2008.

O'Connell, D.P. "Innocent Passage of Warships." 7 *Thesaurus Acroasium, The Law of the Sea* (4th session: September 1976), 1977, 408–51.

Ojanen, Hanna. "Finland: Rediscovering Its Nordic Neighbours After an EU Honeymoon?" *Security Dialogue* 36, No. 3 (2005) 407–11.

Orwell, George. *Nineteen Eighty-Four: A Novel*. London: Secker & Warburg, 1949.

OSCE Document Library. "Open Skies Treaty".

Österlund, Bo. "Changing Scenarios in the Baltic Security Policy From the Historical Perspective." *Journal of East-West Business* 19, No. 1–2 (2013) 63–78.

Ó Tuathail, Gearóid, and Simon Dalby. "Introduction: Rethinking Geopolitics—Towards a Critical Geopolitics." In *Rethinking Geopolitics* (first published London: Routledge 1998) 1–15.

Paasikivi, J.K. "New Year Speech in the New Year of 1950".

Palosaari, Teemu. *The Art of Adaptation: A Study on the Europeanization of Finland's Foreign and Security Policy*. TAPRI Studies in Peace and Conflict Research. Tampere: Tampere University Press, 2011.

Parliament of Åland. "Minutes of the Plenary Session on 3 December 2013." 2013.

——— "Minutes of the Plenary Session on 4 December 2013." 2013.

Petrovsky, Vladimir. "Towards Comprehensive Security Through the Enhancement of the Role of the United Nations (Aide-Mémoire)." *Alternatives: Global, Local, Political* 15, No. 2 (1990) 241–45.

Pharand, Donat. "International Straits." *Thesaurus Acroasium, The Law of the Sea (4th Session September 1976)* 7 (1977) 64–100.

Pisoni, Johan. "Bara var tredje vill att Sverige går med i Nato." *SVT Nyheter*, 3 July 2017.

Poullie, Yannick. "Åland's Demilitarisation and Neutralisation at the End of the Cold War—Parliamentary Discussions in Åland and Finland 1988–1995." *International Journal on Minority and Group Rights* 23, No. 2 (2016) 179–210.

Prezelj, Iztok. "Comprehensive Security and Some Implemental Limits." *Information & Security* 33, No. 1 (2015) 13–34.

Prime Minister's Office Finland. "*Turvallisuus Muuttuvassa Maailmassa* [Security in a Changing World]." 1995.

——— "*Euroopan turvallisuuskehitys ja Suomen puolustus* [European Security Development and the Finnish Defence]." 1997.

——— "Finnish Security and Defence Policy 2004 Government Report 6/2004." 2004.

——— "Government Report on Finnish Foreign and Security Policy 6/2016." 2016.

——— "Government's Defence Report VNS 3/2017." *Government Report 5/2017*, 2017.

Prinsen, Gerard, and Séverine Blaise. "An Emerging 'Islandian' Sovereignty of Non-Self-Governing Islands." *International Journal* 72, No. 1 (2017) 56–78.

"Protocol on the Concerns of the Irish People on the Treaty of Lisbon." *Official Journal of the European Union*, 2 March 2013.

Pynnöniemi, Katri, and Charly Salonius-Pasternak. "Security in the Baltic Sea Region: Activation of Risk Potential." *FIIA Briefing Paper* No. 196 (2016).

Rainne, Juha. *Legal Implications of NATO Membership: Focus on Finland and Five Allied States*. Helsinki: Erik Castrén Research Reports, 2008.

Read, Oliver. "How the 2010 Attack on Google Changed the US Government's Threat Perception of Economic Cyber Espionage." In *Cyberspace and International Relations: Theory, Prospects and Challenges*, edited by Jan-Frederik Kremer and Benedikt Müller. Berlin and Heidelberg: Springer, 2014, 203–30.

Reynolds, Michael A. *Shattering Empires: The Clash and Collapse of the Ottoman and Russian Empires 1908–1918*. Cambridge and New York: Cambridge University Press, 2011.

Roach, J. Ashley, and Robert W. Smith. *Excessive Maritime Claims*. 3rd edn. Leiden: Martinus Nijhoff Publishers, 2012.

Robins, Graham, Håkan Skogsjö, and Jerker Örjans. *Bomarsund: Det ryska imperiets utpost i väster*. Mariehamn: Skogsjömedia, 2004.

Rosas, Allan. "The Åland Islands as a Demilitarised and Neutralised Zone." In *Autonomy and Demilitarisation in International Law: The Åland Islands in Changing Europe*, edited by Lauri Hannikainen and Frank Horn. The Hague and Cambridge: Kluwer Law International, 1997, 23–40.

Rotkirch, Holger. "The Demilitarization and Neutralization of the Aland Islands: A Regime 'in European Interests' Withstanding Changing Circumstances." *Journal of Peace Research* 23, No. 4 (1986) 357–76.

——— "A Peace Institute on the War-Path: The Application of the Treaty on Open Skies to the Neutralized and Demilitarized Åland Islands and the Powers of the Åland Autonomy." In *Nordic Cosmopolitanism—Essays in International Law for Martti Koskenniemi*, edited by Jarna Petman and Jan Klabbers. Leiden: Martinus Nijhoff Publishers, 2003, 61–88.

Schofield, Julian. *Militarization and War*. New York: Palgrave Macmillan, 2007.

Scott, James Brown (ed.). *Argument of the Honorable Elihu Root on Behalf of the United States Before the North Atlantic Coast Fisheries Arbitration Tribunal at the Hague, 1910*, with introduction and appendix by James Brown Scott. Boston: The World Peace Foundation, 1912.

Shabtai, Rosenne (ed.). "League of Nations." *Conference for Codification of International Law* (1930), 1975.

Shaw, Martin. "Twenty-First Century Militarism: A Historical-Sociological Framework." In *Militarism and International Relations: Political Economy, Security, Theory*, edited by Anna Stavrianakis and Jan Selby. Abingdon: Routledge, 2012.

Simón, Luis. *Geopolitical Change, Grand Strategy and European Security*. Basingstoke: Palgrave Macmillan, 2013.

Sipilä, Juha. "*Seinäjoen puoluekokous poliittinen linjapuhe* [Political Speech at Seinäjoki Party Conference]." 11 June 2016.

Société des Nations. "Conférence Relative à la Non-Fortification et à la Neutralisation des Iles d'Aland." *Actes de la conférence*, 1921.

Söderhjelm, J.O. *Démilitarisation et neutralisation des Iles d'Aland en 1856 et 1921*. Helsingfors, 1928.

Spiliopoulou Åkermark, Sia. "Shifts in the Multiple Justifications of Minority Protection." *EYMI* 7 (2010) 5–18.

——— *The Åland Example and Its Components: Relevance for International Conflict Resolution*. Mariehamn: The Åland Islands Peace Institute, 2011.

——— "Åland's Demilitarisation and Neutralisation: Continuity and Change." In *The Åland Example and Its Components: Relevance for International Conflict Resolution*. Mariehamn: The Åland Islands Peace Institute, 2011, 50–71.

———. "The Meaning of Airspace Sovereignty Today—A Case Study on Demiltarisation and Functional Airspace Blocks." *Nordic Journal of International Law* 86, No. 1 (2017) 91–117.

——— "The Puzzle of Collective Self-Defence: Dangerous Fragmentation or a Window of Opportunity? An Analysis With Finland and the Åland Islands as a Case Study." *Journal of Conflict and Security Law* 22, No. 2 (2017) 249–74.

——— "Old Rules and New Technology: Drones and the Demilitarisation and Neutralisation of the Åland Islands" (forthcoming).

Spiliopoulou Åkermark, Sia, Holiencin, Filip, and Timo Koivurova. "Untying a Sailor Knot" (forthcoming).

Spingarn, Jerome H. "Five Months in London." *Bulletin of the Atomic Scientists* 13, No. 7 (September 1957).

Stahn, Carsten. *The Law and Practice of International Territorial Administration: Versailles to Iraq and Beyond*. Cambridge: Cambridge University Press, 2010, 257–61.

Stavrianakis, Anna. "Legitimising Liberal Militarism: Politics, Law and War in the Arms Trade Treaty." *Third World Quarterly* 37, No. 5 (2016) 840–65.

Stavrianakis, Anna, and Jan Selby. *Militarism and International Relations: Political Economy, Security, Theory*. London: Routledge, 2013.

Stearns, Peter N. *Demilitarization in the Contemporary World*. Urbana: University of Illinois Press, 2013.

Stephan, Sarah. "Making Autonomies Matter: Sub-State Actor Accommodation in the Nordic Council and the Nordic Council of Ministers—An Analysis of the

Institutional Framework for Accommodating the Faroe Islands, Greenland and Åland Within 'Norden'." *European Diversity and Autonomy Papers (EDAP)*, No. 3 (2014).
——— "Självstyrelse och Regional Integration—Nordiska lösningar." In *Självstyrelser i Norden i ett fredsperspektiv—Färöarna, Grönland och Åland*, edited by Sia Spiliopoulou Åkermark and Gunilla Herolf. Mariehamn: Nordiska rådet & Ålands fredsinstitut, 2015, 78–87.
Stern, Eric K. "Bringing the Environment in: The Case for Comprehensive Security." *Cooperation and Conflict* 30, No. 3 (1995) 211–37.
Stjernfelt, Bertil. *Ålands hav och öar—brygga eller barriär? Svensk-finsk försvarsfråga 1915–1945*. Stockholm: Marinlitteraturföreningen, 1991.
Stråth, Bo. *Europe's Utopias of Peace*. London: Bloomsbury, 2016.
Swedish Government. "Statens Offentliga Utredningar, SOU 2012:48 'Maritim samverkan'." 2012.
Tanaka, Yoshifumi. *The International Law of the Sea*. 2nd edn. Cambridge: Cambridge University Press, 2015.
"Territorial Surveillance Act 755/200 and Amendments in 195/6.3.2015".
Thorhallsson, Baldur, and Alyson J.K. Bailes. "Do Small States Need 'Alliance Shelter'? Scotland and the Nordic Nations." In *Security in a Small Nation: Scotland, Democracy, Politics*, edited by Andrew W. Neal. Cambridge, UK: Open Book Publishers, 2017, 49–75.
Thorhallsson, Baldur, and Anders Wivel. "Small States in the European Union: What Do We Know and What Would We Like to Know?" *Cambridge Review of International Affairs* 19, No. 4 (2006) 651–68.
Tiilikainen, Teija. *The Åland Islands, Finland and European Security*. Mariehamn: Åland Islands Peace Institute, 2002.
——— "Åland in European Security Policy." In *The Nordic Countries and the European Security and Defence Policy*, edited by Alyson J.K. Bailes, Gunilla Herolf and Bengt Sundelius. Oxford: Oxford University Press, 2006.
Tollet, Artur, and John Uggla. *Lagstiftningen angående självstyrelse för Åland: Jämte tillhörandeförfattningar*. Helsingfors: Holger Schildt, 1930.
Törnudd, Klaus. *Soviet Attitudes Towards Non-Military Regional Co-Operation*. Vol. 28:1. Helsingfors: Societas Scientiarium Fennica. Commentationes Humanarum Litterarum, 1961.
"Treaty of Dorpat Between Finland and Russia." *LNTS* Vol. 3 (1921), Issue 1, No. 91.
"Treaty of Lisbon." *OJ* C 306 (2007).
Tsagourias, Nikolaos, and Russell Buchan. *Research Handbook on International Law and Cyberspace*. Cheltenham, UK: Edward Elgar Publishing, 2015.
Tudeer, Erik. "Ålands hemvärn 1939–1940." *Särtryck ur krigshistorisk tidskrift*, No. 9 (1990).
Tuomioja, Erkki. "Minister of Foreign Affairs in a Seminar *'Åland och demilitarisering i dag'*" on 7 March 2005 in Mariehamn Åland.
Ullman, Richard. "Redefining Security." *International Security* 8, No. 1 (1983) 129–53.
UN (First) United Nations Conference on the Law of the Sea, Official Records, Volume III.

UN GA Res. A/RES/41/92 (4 December 1986).

UN GA Res. A/42/574 (18 September 1987).

"Uniform Interpretation of Norms of International Law Governing Innocent Passage." *Law of the Sea Bulletin* 12, No. 14 (1989).

UN Third United Nations Conference on the Law of the Sea, Official Records, Volume II, Volume III, Volume XIV, Volume XVI, Volume XVII.

US Department of Defense. "Freedom of Navigation." *Report for Fiscal Year 2013.*

US Department of State. "Limits in the Seas No. 112." *United States Responses to Excessive Maritime Claims*, 1992.

Vagts, Alfred. *A History of Militarism: Civilian and Military*. New York: The Free Press, 1967.

Vanhanen, Matti. "*Pääministerin ilmoitus eduskunnalle EU:n hallitustenvälisen konferenssin johdosta* [Prime Minister's Announcement Made to the Parliament Concerning the Intergovernmental Conference of the EU] on 22 June 2004." 2004.

Venice Commission (Council of Europe). "Report on the Democratic Control of the Armed Forces." 2008.

Verzijl, Jan Hendrik Willem. *International Law in Historical Perspective: Volume 3, State Territory*. Leyden: Sijthoff, 1970.

——— *International Law in Historical Perspective. Volume 10, The Law of Neutrality*. Leyden: Sijthoff, 1979.

Visuri, Pekka. *Puolustusvoimat kylmässä sodassa: Suomen puolustuspolitiikka vuosina 1945–1961*. Porvoo: Söderström, 1994.

von Clausewitz, Karl. *Hinterlassene Werke über Krieg und Kriegführung*. Berlin: Ferdinand Dümmler, 1832–1837.

von Mohl, Robert. *Die Polizei-wissenschaft nach den Grundsätzen des Rechtsstaates*. 2nd edn., Tübingen: H. Laupp, 1844.

Vuorela, Heikki. "Nato-jäsenyyden vastustus väheni—Kanerva yllättyi muutoksen vähäisyydestä." *Maaseudun Tulevaisuus*, 11 January 2017.

Vuori, Juha A. *How to Do Security With Words: A Grammar of Securitisation in the People's Republic of China*. Turku: University of Turku, 2011.

Waever, Ole. "Securitization and Desecuritization." In *On Security*, edited by Ronnie D. Lipschutz. New York: Columbia University Press, 1995.

——— "The EU as a Security Actor: Reflections From a Pessimistic Constructivist on Post-Sovereign Security Orders." In *International Relations Theory and the Politics of European Integration: Power, Security and Community*, edited by Morten Kelstrup and Michael C. Williams. London: Routledge, 2000.

Wahlbäck, Krister. *Finlandsfrågan i svensk politik 1937–1940*. Stockholm: P. A. Norstedt och söners förlag, 1964.

Wahlberg, Peter (ed.). *International Treaties and Documents Concerning the Åland Islands 1856–1992*. Mariehamn: Ålands kulturstiftelse, 1993.

Waldman, Thomas. *War, Clausewitz and the Trinity*. Farnham & Surrey: Ashgate, 2013.

Weber, Max (Peter Lassman, and Ronald Speirs eds.). *Weber: Political Writings*. Cambridge: Cambridge University Press, 1994.

Weiss, Tomáš. "The Blurring Border Between the Police and the Military: A Debate Without Foundations." *Cooperation and Conflict* 46, No. 3 (2011) 396–405.

Westing, Arthur H. "The Environmental Component of Comprehensive Security." *Security Dialogue* 20, No. 2 (1989) 129–34.

Wigell, Mikael, and Antto Vihma. "Geopolitics Versus Geoeconomics: The Case of Russia's Geostrategy and Its Effects on the EU." *International Affairs* 92, No. 3 (2016) 605–27.

Wilde, Ralph. *International Territorial Administration: How Trusteeship and the Civilizing Mission Never Went Away*. Oxford and New York: Oxford University Press, 2008.

Yang, Haijiang. *Jurisdiction of the Coastal State Over Foreign Merchant Ships in Internal Waters and the Territorial Sea*. Berlin: Springer, 2006.

"*Yearbook of the International Law Commission*." Volume I (1954).

Zagoria, Donald S. Bookreview of "Beyond War: Japan's Concept of Comprehensive National Security." *Foreign Affairs* 64, No. 1 (1985) 193.

Index

For Product Safety Concerns and Information please contact our EU
representative GPSR@taylorandfrancis.com
Taylor & Francis Verlag GmbH, Kaufingerstraße 24, 80331 München, Germany

www.ingramcontent.com/pod-product-compliance
Ingram Content Group UK Ltd.
Pitfield, Milton Keynes, MK11 3LW, UK
UKHW022157190926
475702UK00003B/71
9781138093300